Möllerke · Im Ausland unterwegs

# Im Ausland unterwegs

Technicians and Engineers Abroad

Georg Möllerke

Umschlagillustration: ZEFA/STOCKMARKET
TSM 7388; Photo: Mug Shots


Die Deutsche Bibliothek - CIP-Einheitsaufnahme

**Moellerke, Georg:**
Im Ausland unterwegs : technicians and engineers abroad /
Georg Moellerke. - 2. Aufl. - Düsseldorf : VDI-Verl., 1996
ISBN 978-3-540-62357-1    ISBN 978-3-642-48669-2 (eBook)
DOI 10.1007/978-3-642-48669-2

# Vorwort

Auf Auslandsreisen kann man als Ingenieur oder Techniker meistens wegen des Umfanges kein Fachlehrbuch für die englische Sprache mitnehmen, wohl aber einer handlichen Sprachführer für technisches Englisch, ein Taschenbuch mit Redewendungen und Gesprächen also, Englisch-Deutsch, wie dieses: „Im Ausland unterwegs - Technicians and Engineers Abroad".

Wem ist es nicht schon ähnlich ergangen: Wir sprechen mit Fachleuten im Ausland und stellen fest, daß uns gewisse ganz alltägliche Ausdrücke der Umgangssprache einfach nicht einfallen wollen oder gar unbekannt sind. Dabei sind einem die reinen Fachausdrücke wohl geläufig. Hier will dieses Buch helfen.

In dem Taschenbuch sind die Gespräche nach Situationen zusammengestellt. So findet man, entsprechend den gegebenen Umständen, in den meisten Fällen die benötigten Sätze, oder diese lassen sich hiervon ableiten.

Im Flugzeug, vor einer Landung in New York zum Beispiel, kann man sich Gesprächssituationen vor Augen führen, denen man in den nächsten Stunden begegnen wird: Ich treffe einen Kunden zum ersten Mal: der Geschäftspartner erscheint verspätet auf dem Flughafen; ich treffe mit Fachkollegen auf dem Prüffeld vor einer wichtigen Maschinen- oder Geräteabnahme zusammen; ein Bauteil ist während der Ferigung zu besichtigen; ein Monteur berichtet über eine Anlagenstörung. Neben den fachlich orentierten Abschnitten sind allgemeine Redewendungen für den „Reise-Alltag" aufgeführt.

Das Kernstück bilden die neun Episoden de Serie "The experiences of Bob Keller, an engineer". Auf kurzweilige Art werden darin die Erlebnisse eines Ingenieurs auf seiner Kanada-Reise geschildert.

Die Sätze der einzelnen Abschnitte sind ins Deutsche übersetzt, so daß „Im Ausland unterwegs - Technicians and Engineers Abroad" auch von englischsprachigen Fachleuten benutzt werden kann.

Ich hoffe sehr, daß auch Sie recht viel Nutzen aus diesem Taschenbuch ziehen können.

Nussbaumen/Schweiz, im März 1996          Ing. *Georg Möllerke*

# Contents

# The experiences of Bob Keller, an engineer

The material presented in this series relates events in the working life of Robert Keller (known as Bob), section leader of Suter Engines Ltd., Aarau. The various episodes deal with typical situations in his day-to-day work, such as visiting customers abroad, discussions with laboratory personnel, attending works tests and conversations with designers and erecting engineers.

## Episode 1

### On a business trip to Toronto

**Bob's flight Air Canada No. 645 will be landing shortly at the International Airport in Toronto.**

Announce-
ment:
In a few minutes we shall be landing in Toronto. Please refrain from smoking and fasten your seatbelts.
*After landing all passengers head for the gates designated ‹Passport Control›. Bob Keller presents his passport to the official at the gate. After examining the document the official indicates that he may pass. Bob now walks to the customs gate.*

Customs:
Do you have anything to declare, sir?

Bob Keller:
No, I have only my luggage with my personal belongings with me.

Customs:
Good. Enjoy your stay, sir.
*Eugene Randall, the area representative of Suter Engines Ltd., is waiting for Bob in the arrival hall. Eugene has spotted Bob and waves to him from a distance. They finally meet in the hall.*

Eugene:     Hello Bob. Have you had a pleasant flight?
Bob:        Yes indeed, I had a very good trip.
            *After a short talk on business matters.*
Eugene:     **Please come along to my car. I'll drive you straight to
            the hotel. Our secretary, Brenda Carlsson, made a reser-
            vation for you at the Hotel Continental. It's located in
            downtown Toronto. The ride will take us about three
            quarters of an hour.**
            *They arrive at the hotel.*
Bob:        I remember this part of the city.
Eugene:     Here we are, Bob.
            *Eugene opens the rear trunk of his car and both carry
            the suitcases to the reception hall of the hotel.*
Bob:        Looks rather expensive.
Eugene:     That's quite all right.
            *They go to the reception desk.*


**experiences** Erfahrungen, Erlebnisse – **to present** aufzeigen, darstellen, präsentieren – **to relate** berichten über, erzählen – **event** Vorfall, Ereignis – **section leader** Gruppenleiter – **engine** Verbrennungs(kraft)-maschine; hier gemeint: (Diesel-)Motor – **various** verschieden – **to deal with** handeln von – **to visit** besuchen – **customer** Kunde – **abroad** im Ausland – **laboratory personnel** Laborpersonal – **to attend** beiwohnen, dabei sein – **designer** Konstrukteur – **erecting engineer** Montage-Ingenieur, Monteur – **business trip** Geschäftsreise – **will be landing shortly** wird gleich landen – **to refrain from** einstellen, Abstand nehmen von – **to fasten seatbelts** Gurte anlegen – **to head for** strömen, gehen zu – **gate** Einlasspforte – **designated** bezeichnet – **to present** vorzeigen – **official** Beamter – **to examine** überprüfen – **to indicate** andeuten; sonst auch: zeigen – **customs** Zoll – **to declare** verzollen, deklarieren – **luggage** Gepäck – **enjoy your stay** (ich wünsche) schönen Aufenthalt – **area representative** Gebietsvertreter – **arrival hall** Ankunftshalle – **to spot** erkennen, ausmachen – **final(ly)** schliesslich – **located** gelegen – **downtown** mitten in der Stadt, in der City – **ride** hier: Fahrt – **to arrive** ankommen, eintreffen – **part** Teil – **rear trunk** hinterer Kofferraum – **suitcase** Koffer – **reception desk** Empfang(spult)

## Episode 2

# At the Continental Hotel, Toronto

**Bob and Eugene have arrived at the reception desk and are being received by the receptionist.**

| | |
|---|---|
| Eugene: | We have made a reservation for Mr. Robert Keller of Suter Engines. |
| Recept.: | That's all right. Please fill out our check-in card. Thank you. Here's your key. You have room number 343, on the third floor to the left of the elevator. |
| Bob: | Thank you. |
| Eugene: | Well, that's settled. |
| Bob: | When do you think we could meet tomorrow? |
| Eugene: | I'd like to pick you up at nine o'clock. |
| Bob: | Would suit me fine. |
| Eugene: | At ten o'clock we are meeting Mr. Watson, the head of Technical Projects and Equipment Purchasing at the Post Office of Toronto. |
| Bob: | I think we have a great chance of winning the order for the emergency diesel generating set. |
| Eugene: | Yes, the chances appear to be very good. With a bit of luck we could be rewarded with the contract by the end of August. |
| Bob: | That would be fine. By the way, I have incorporated some improvements in the cooling water system. This will reduce the lump sum price considerably. |
| Eugene: | By how much? |
| Bob: | I can't say for definite yet. I must do a few calculations in my hotel room. |
| Eugene: | Wouldn't it be better for us to meet a bit earlier, let's say at eight thirty? |
| Bob: | I quite agree. I could then explain the modified water system to you before we go to the meeting. |
| Eugene: | But now I'll go. You probably are looking forward to a rest. Have a nice time. So long, Bob. |
| Bob: | Good night then, see you at eight thirty tomorrow morning. |

**to arrive** ankommen, eintreffen – **reception desk** Empfang(spult) –**to receive** empfangen – **to fill out** (or:in) ausfüllen – **check-in card** Anmeldeformular – **elevator** (US) Fahrstuhl, Lift – **that's settled** das wäre erledigt – **to pick up** abholen – **to suit** passen – **head** Chef, Leiter – **purchasing** Einkauf – **order** Auftrag – **emergency diesel generating set** Notdieselaggregat – **to appear** scheinen zu sein – **to reward** zuerkennen, gewinnen – **by the way** übrigens – **to incorporate** vorsehen, einschliessen – **improvement** Verbesserung – **to reduce** herabsetzen, reduzieren – **lump sum price** Gesamtpreis (der Offerte) – **considerable (-bly)** beträchtlich – **by how much?** um wieviel? – **to agree** einverstanden sein, zustimmen – **to explain** erklären – **modified** modifiziert, abgeändert – **to look forward to** sich freuen auf

**MinProg activity**

**The experts of the Ministry of Progress had no problem in creating this phantastic 'fly', but they're having trouble finding a suitable name for it. 'Well, I propose Cruxi or ...'**

## Episode 3

# Preparing for a meeting with Mr. Watson of the Toronto Post Office

**Before Bob and Eugene start negotiations with Mr. Watson they study the details of their tender.**

Eugene: What's the exact rating of the emergency diesel-driven generating set?

Bob: It's rated at 900 kW, 440 V, 60 Hz, 1800 rev/min. Here's the drawing I wanted to show you. This part covers the modified cooling water system.

Eugene: It's simpler than I first thought it would be. It will surely save us some money; we'll be very competitive with your system.

Bob: As we already have supplied five sets for the Toronto PTT we are naturally in a good position.

Eugene: Indeed, we are. Please show me the diagram with the automatic starting facilities.

Bob: This set operates up on failure of the mains supply as follows: The signal transferred from auxiliary contact S5 to ...
(Bob explains the system to Eugene.)

Eugene: There are some improvements on our starting system, too. It's much simpler to operate now.

Bob: That's true. Especially with regard to test starting. You may follow each step until the whole sequence has been through.

Eugene: Very well, Bob. Just a few words on the generator itself. *(After Bob's describing the generator, they leave for the meeting with Mr. Watson of the Toronto Post Office.)*

**to prepare** (sich) vorbereiten – **negotiation** Verhandlung – **details** Einzelheiten – **tender** Offerte mit Spezifikation – **rating** Auslegung; auch: Leistung, Bemessung – **rated at** ausgelegt, bemessen mit – **rev/min** revolutions per minute: Umdrehungen pro Minute – **drawing**

Zeichnung – **part** Teil – **to cover** umfassen, abdecken – **modified** modifiziert – **to save** (ein)sparen – **competitive** konkurrenzfähig – **to supply** liefern – **set** Satz, Aggregat – **starting facilities** Starteinrichtungen – **to operate up on** hier: hochlaufen (oder: starten) bei – **failure** Ausfall – **mains supply** Netzversorgung – **to transfer** überleiten, weiterleiten – **auxiliary contact** Hilfskontakt (an einem Relais) – **with regard to** hinsichtlich – **step** Stufe – **sequence** Folge(ablauf) – **to describe** beschreiben

**There's nothing like the work simulator designed by the Ministry of Progress.**

Episode 4

# Meeting with Mr. Watson of the PTT technical division

**Bob and Eugene discuss the tender concerning an emergency set for the Toronto Post Office.**


| | |
|---|---|
| Recept.: | Good morning, gentlemen. What can I do for you? |
| Eugene: | We are Randall and Keller from Suter Engines Ltd. We have an appointment with Mr. Watson at ten o'clock. |
| Recept.: | Oh yes. You are from Aarau, Switzerland, aren't you? |
| Bob: | That's right. I'm from Aarau. Mr. Randall is the area representative of Suter Engines. |
| Recept.: | We've been expecting you. I'll take you up to Mr. Watson straight away. |
| Eugene: | Thank you.<br>(In Mr. Watson's office.) |
| Watson: | Hello Mr. Keller, Mr. Randall. Nice to see you again.<br>(After some general talk.)<br>Gentlemen, I'm really sorry, but there's something in your tender which is posing us a problem. |
| Bob: | Oh! Which point do you mean, Mr. Watson? |
| Watson: | The space available in the Post Office basement, where the diesel generating set will be installed, is not sufficient to get your set in. Have a look at this drawing. |
| Eugene: | What's the space shortage? |
| Watson: | The basement is 0.5 m too short to accept the overall length of your set. |
| Bob: | Well, I see a possibility to solve this problem. |
| Watson: | But how, Mr. Keller? |
| Bob: | We've quoted an in-line engine. We can just as easily provide our 'V' type engine. There'll be sufficient space then. This engine is 0.8 m shorter than our in-line design. |
| Eugene: | Bob, you've drawings of the 'V' type engines with you. Let's investigate at once. |
| Watson: | I can tell you, I would be very happy if the problem could be solved by using a 'V' type engine. Let's continue! |

**division** hier: (Geschäfts-)Bereich – **to discuss** diskutieren, erörtern – **tender** Offerte mit Spezifikation – **concerning** betreffend – **appointment** Verabredung – **area representative** Gebietsvertreter (Leiter) – **to expect** erwarten – **to take up** hinaufbringen – **straight away** sofort – **to pose** (Problem) aufwerfen – **space** Platz, Raum – **available** zur Verfügung – **basement** Kellergeschoss; auch: Grundmauern – **sufficient** ausreichend – **space shortage** fehlender Platz – **to accept** hier: aufnehmen – **to solve** lösen – **to quote** Preisofferte machen – **in-line engine** Reihenmotor – **to provide** vorsehen, liefern – **'V' type engine** V-Motor – **design** Konstruktion, Ausführung – **to investigate** untersuchen – **to continue** weitermachen, fortfahren

**There's been an alarming increase in the number of things I know nothing about.**

# Episode 5

## Mr. Watson makes his decision in favour of the Suter 'V' engine

**Having discussed the engine/generator tender in detail, Mr. Watson of the Toronto PTT has no further objections to the Suter set. There also seems to be sufficient space for installing the equipment in the Post Office basement.**

| | |
|---|---|
| Watson: | When comparing your tender with the other four competitors, I found that yours was surprisingly moderate in price. How's that possible? |
| Bob: | We've included a few improvements, such as the newly designed cylinder head and the 'Superjet' injection pump which allowed us to maintain product quality for lower production costs. |
| Eugene: | We also have a simplified cooling water system; simplified as to maintenance, too. |
| Watson: | I must admit, we've never had any serious trouble with any of your sets installed in our premises. You may expect our written order in about four weeks time. |
| Bob: | Thank you, Mr. Watson. We always enjoyed a good relationship with the Toronto PTT. |
| Watson: | I would be happy if we could talk about a preliminary progress schedule. You needn't give me any definite dates now. |
| Bob: | Certainly, we can give you some data in advance. You'll receive the definite schedule in approximately two weeks from now. As you will understand, I'll have to talk over various details with the manufacturing division. |
| Watson: | That's quite all right. However, what I really need as quickly as possible are the confirmed data of the 'V' engine, also of the base plate. Well, have we dealt with everything? |
| Eugene: | I think so. Mr. Watson, if you think of anything else, simply ring me up. You have the new address of my Toronto office, haven't you? |

Watson:     Yes, I do have your new address. Then, gentlemen, I
            don't need to keep you any longer. Many thanks for
            your comprehensive information, it's really a pleasure
            working with people who know their business.


**decision** Entscheidung – **in favour of** zugunsten von – **objection**
Einwand – **to compare** vergleichen – **competitor** Konkurrent – **surprising (ly)** überraschend – **moderate** niedrig, gemässigt – **to include**
einbringen, einschliessen – **to design** konstruieren, entwerfen –
**cylinder head** Zylinderkopf – **injection pump** Einspritzpumpe – **to
maintain** beibehalten – **maintenance** Wartung – **to admit** zugeben,
zugestehen – **serious** ernst(haft) – **premises** (Pl.) Bauten,
Räumlichkeiten – **relationship** Beziehung – **preliminary** vorläufig,
Vorläufer ... – **progress schedule** Fabrikationsprogramm – **in advance**
im voraus – **schedule** Aufstellung, Liste – **various** verschieden –
**manufacturing** Fabrikations ... – **confirmed** bestätigt – **base plate**
Grundplatte – **have we dealt with everything?** haben wir alles
besprochen? – **to keep** hier: aufhalten – **comprehensive** umfassend

**What would you do if it really would hatch?**
**(to hatch: ausschlüpfen)**

Episode 6

## Mr. Fisher of the Canadian Overseas Shipping Company arrives at the Suter Engines Subsidiary in Toronto

**Mr. Fisher, the technical inspector of COSC is visiting Suter Engines the first time. He meets Eugene and Bob to discuss the purchase of six diesel engines, each rated at 4800 h.p. and equipped with Engine Diagnostic System (EDS).**

| | |
|---|---|
| Fisher: | Good morning. My name is Fisher (after a knock at the door). |
| Brenda: | (Eugene's secretary). Ah yes, Mr. Fisher. Mr. Randall is expecting you. I'll take you in to see Mr. Randall right away. (Brenda enters Eugene's office) Eugene, Mr. Fisher's here. |
| Eugene: | Ah, good morning, Mr. Fisher. How do you do? |
| Fisher: | How do you do? |
| Eugene: | This is Mr. Keller from our Aarau head office. |
| Bob: | How do you do. |
| Eugene: | Please sit down, won't you? Plane was a bit late, wasn't it? |
| Fisher: | Yes, nearly an hour late. There were quite a few planes in the air waiting to land before us.<br>(After some general talk)<br>Well, Mr. Randall, we're very much interested in your large diesel engines equipped with the EDS system. You've already had a considerable amount of experience with this system, haven't you? |
| Eugene: | Yes indeed. We've been equipping our large engines with this system for the last three years. However, this is an advanced design, a modified version of our previous design. We've improved the technique for monitoring the piston ring wear. |
| Bob: | The EDS system proved to be most reliable. It really is a great asset to be able to check the cylinder, piston and piston ring wear at any time, simply by pressing a button on the data logger. |

| | |
|---|---|
| Fisher: | So that it's no longer necessary to dismantle an engine in order to inspect the condition of those parts? |
| Eugene: | Exactly. Thanks to electronics. Look, before going into detail, would you like a cup of coffee—or something to eat? |
| Fisher: | I could do with some coffee. |
| Bob: | Good. Let's go and get some, shall we? |

**subsidiary** Tochtergesellschaft, Filiale – **purchase** Ankauf – **rated at** hier: mit einer (Nenn-)Leistung von – **to equip** ausrüsten, versehen – **engine diagnostic** Motordiagnose … – **to expect** erwarten – **head office** Stammhaus – **plane** Flugzeug – **quite a few** eine Menge (Untertreibung) – **considerable** beträchtlich – **amount** Menge – **experience** Erfahrung – **advanced** weiterentwickelt – **design** Konstruktion – **previous** vorherig – **to improve** verbessern – **to monitor** überwachen – **piston ring wear** Verschleiss an Kolbenringen – **to prove** (sich) erweisen – **reliable** betriebssicher, zuverlässig – **asset** Vorteil, Gewinn – **able** imstande – **to press** drücken – **button** (Druck-)Knopf – **necessary** notwendig – **to dismantle** auseinandernehmen, zerlegen – **condition** Zustand

# Engineering Report

Monatszeitschrift für technisches Englisch
Verlag, Redaktion, Copyright:
Georg Moellerke, El.-Ing. und staatl. geprüfter Übersetzer
Kornweg 5, CH-5415 Nussbaumen (Schweiz)

## Episode 7

## Introducing the electronic EDS system to Mr. Fisher, the shipping company's technical inspector

**Henry Winter, an expert in Engine Diagnostic Systems (EDS), is requested by Eugene and Bob to inform Mr. Fisher about control equipment for large engines. A special item is the measuring technique for monitoring piston ring conditions.**

| | |
|---|---|
| Eugene: | Mr. Fisher, let's go up to the conference room. Our Mr. Winter has prepared a 'lecture' to give you an insight into our monotoring systems. |
| | Brenda, Mr. Fisher, Bob and I are going up to the conference room on the second floor. Just so you know where we are. |
| Brenda: | That's all right, Eugene. |
| | (Entering the conference room) |
| Eugene: | This is Mr. Winter. |
| Fisher: | How do you do. |
| Henry: | How do you do. Will you please sit here, Mr. Fisher. You'll be able to see the slides better from here. I'll give you a general description of the EDS system. |
| | The aim of our diagnostic system is to achieve better operation economy. EDS, a computerised on-line system, continuously monitors the major engine components and indicates, at an early stage, of any abnormal conditions which, if undetected, could lead to damages and consequently to costly repairs. Furthermore, the EDS system analyses and records any change in condition of essential components, informs of possible undesirable developments thus providing the necessary data to help plan overhaul jobs previously carried out at fixed intervals, so avoiding unnecessary dismantling of the diesel engine. The comparatively large number of sensors involved, the amount of data and the complex analysis carried out at high frequency call for a mini-computer based electronic data system ... |
| | (After some time) |

Fisher:     Thank you, Mr. Winter, that certainly was very interesting and a lot of information; I think I ought to digest it before going on.

Eugene:     A good idea, indeed. Let's take a break for some coffee. (Pause)

Henry:      We shall now continue with piston ring monitoring. Piston ring wear is detected by measuring the wear of chromium plating on the rings. A proximity sensor picks up signals from a chromium plated ring, the first ring, and a pure cast ring, here number two. The processor then compares the signals from these two readings and calculates the expected lifetime. (Henry Winter eventually comes to the end of his lecture) I'm now going to show you a typical piston ring arrangement: fire ring, compression rings one, two and three ...

**to introduce** vorführen, einführen – **engine diagnostic** Motorendiagnose... – **to request** bitten, fragen – **control equipment** Steuereinrichtung(en) – **item** Punkt – **measuring technique** Messtechnik – **to monitor** überwachen – **piston ring condition** Zustand der Kolbenringe – **to prepare** vorbereiten – **lecture** Vortrag, Vorlesung – **insight** Einblick – **to enter** betreten, eintreten – **slide** Diapositiv – **description** Beschreibung – **aim** Ziel, Zweck – **to achieve** erreichen – **operation economy** Betriebswirtschaftlichkeit – **continuous(ly)** laufend, fortwährend – **major** wichtigst, Haupt ... – **component** Einzel-Teil – **to indicate** anzeigen – **stage** Stadium – **undetected** unentdeckt – **damage** Schaden – **consequent(ly)** folglich – **to analyse** analysieren – **to record** aufzeichnen – **essential** wichtig, wesentlich – **undesirable** unerwünscht – **development** Entwicklung – **to provide** vorsehen, liefern – **overhaul job** Überholarbeit – **fixed** fest – **interval** Zeitraum – **to avoid** vermeiden – **dismantling** Demontage, Auseinandernehmen – **comparative(ly)** vergleichsweise – **sensor** Fühler – **involved** erforderlich, betroffen – **frequency** Häufigkeit, Frequenz – **to call for** verlangen – **to digest** verdauen – **break** Unterbrechung, Pause – **chromium plating** Chromplattierung – **proximity sensor** Näherungsfühler – **to pick up** aufnehmen – **pure** rein – **cast ring** Gussring – **to compare** vergleichen – **reading** Wert, Ablesung – **eventual(ly)** schliesslich – **arrangement** Anordnung – **fire ring** Feuerring – **compression ring** Kompressionsring

## Episode 8

# An erecting engineer relates his experience in Mexico

**Eugene, Brenda and Bob have a visitor in Eugene's office. David Freeman, the chief erecting engineer, talks about some problems encountered in putting a large diesel engine in operation.**

| | |
|---|---|
| Brenda: | Did you see the note on your desk? |
| Eugene: | No—what note? Oh yes, here it is. David's coming to see us at 10 o'clock. |
| Bob: | David Freeman, the chief erecting engineer? |
| Brenda: | Quite right. David Freeman. He wants to report on his trip to Mexico. On the telephone he said he had an interesting story for us. |
| Eugene: | Whenever David goes abroad he always comes back with some thrilling story or other. |
| Brenda: | You're telling me! I'll never forget that cliffhanger about his emergency landing in the African jungle. Nobody can tell us there isn't any adventure these days. |
| Eugene: | No, indeed. Not when you think of David. (A knock at the door) |
| David: | Good morning, Brenda. Good morning, Eugene. Hello Bob, how are you? (After some general talk) |
| Eugene: | Thanks for the telex—it was good to hear you'd got the diesel engine running properly. And nice to have you back so soon. I'm glad it only took four days. |
| David: | Well, the actual work only took me five hours. |
| Brenda: | Tell us all about it. |
| David: | Well, it's an odd story. The manager of the power plant told me the diesel engine had been installed according to factory instructions, but it would't start. When I suggested trying to start it up, he said, 'If you can get that diesel engine going, we'll give you four weeks holiday in Mexico—all expenses paid.' |
| Bob: | Quite a challenge! |
| Eugene: | Yes—though I'm sure there isn't any diesel engine you can't put into operation. Well—go on. |

David:      I barred the engine by hand and checked that valve and pump action were both in order. After that, I knew the problem must be something simple. But what?

Bob:      Perhaps the fuel supply?

David:      No, everything seemed to be in order. Then I cranked the engine with the air starting valve. She began to roll, but stopped after a few turns. Then I asked the manager for a long extension ladder to reach the roof. He couldn't see why, in fact he said, 'What we really need is a stick of dynamite to get this going.' Well, I put the ladder up against the exhaust pipe, which extended through the roof. And then I climbed up to the 10-inch connecting flange just under the roof where the muffler was mounted.

Brenda:      I suppose you already knew what was wrong?

David:      Well, I had an idea. When the engine had first started firing, I'd noticed a small whiff of smoke escaping from the flanged joint. So I asked the mechanic to remove the steel blanking flange from between the pipe flanges. It hadn't been removed when the unit was installed.

Bob:      Did the engine start then?

David:      Yes, this time it started without any trouble.

All:      Bravo. Congratulations.

**erecting engineer** Montage-Ingenieur, Monteur – **to relate** berichten, erzählen (von, über) – **visitor** Besucher – **to encounter** stossen auf, begegnen – **to put in operation** in Betrieb setzen – **to report** berichten – **thrilling** spannend, aufregend – **cliffhanger** spannende Geschichte; wörtlich: von jemandem, der an einer Klippe hängt – **emergency landing** Notlandung – **adventure** Abenteuer – **proper(ly)** ordnungsgemäss – **actual** wirklich, tatsächlich – **power plant** Kraftwerk – **factory instructions** Anweisungen des Herstellers – **expenses** Ausgaben – **challenge** Herausforderung – **to bar** (durch)drehen – **valve action** Arbeitsweise der Ventile – **fuel supply** Brennstoffzufuhr – **to crank** (durch)drehen – **extension ladder** Ausziehleiter – **roof** Dach – **stick of dynamite** Stück (oder: Paket) Dynamit – **exhaust pipe** Auspuffrohr – **to climb up** hinaufsteigen – **connecting flange** Anschlussflansch – **muffler** Geräuschdämpfer – **to mount** unterbringen, montieren – **to fire** zünden – **whiff of smoke** kleine Rauchwolke – **to escape** hervorkommen, entkommen – **flanged joint** Flanschverbindung – **steel blanking flange** Blindflansch aus Stahl (Scheibe zum Absperren einer Rohrleitung)

## Episode 9

## Shortly before Bob leaves Toronto for Aarau (via Zurich)

**Bob and Eugene have tackled all problems assigned to them. Bob is ready for his home flight to Zurich.**
**They recall their experiences, especially with regard to reception of the Engine Diagnostic System (EDS). Now, with an attractive order in his pocket, Bob is looking forward to resuming his work at the home office.**

| | |
|---|---|
| Bob: | It's amazing how quickly the days have gone by since I joined you here in Toronto. |
| Eugene: | Time flies. But on the whole we can be satisfied with our achievements. |
| Bob: | It was a rewarding visit, indeed. Think only of our talk with Mr. Fisher, the Shipping Company's inspector. I do believe that there will be a specific enquiry for our EDS system for his ships. |
| Eugene: | You may be right. He was really enthusiastic when we explained the wear control system to him. |
| Bob: | And Mr. Watson was impressed when we handed him the base plate drawing for the 900 kW diesel generating set. He was expecting it a week later. |
| Eugene: | He was certainly happy to receive it earlier. |
| Eugene: | What do you think, Bob, shall I keep you personally up to date on the prospective PTT contract, or will the official information be sufficient for you? |
| Bob: | The normal routine mail will do. Only if something out of the ordinary turns up, then please send me a personal telex. |
| Eugene: | In an urgent case I'll ring you up, of course. |
| Bob: | That will be fine. |
| Eugene: | Coming back to the EDS system. The slides are not bad. But when will the educational film be available? |
| Bob: | By the end of September, I hope. |
| Eugene: | That should be all right. There seems to be a good market potential for our EDS system, doesn't it? |

Bob:     I think so, too. Oh, Eugene, it's nearly 12 o'clock, I must leave now. Have a nice time, we'll meet again in October, won't we?

Eugene:     Yes, at the latest.

Bob:     And many thanks, Eugene, for helping me so much in Toronto. We had an enjoyable and successful time together.

Announcement:     Swissair flight number 902, passengers to Zurich are requested to go to gate 14.

**to leave** abreisen von, verlassen – **to tackle** fertigwerden mit, behandeln – **to assign** zuweisen, zuordnen – **with regard to** hinsichtlich – **reception** Aufnahme, Anerkennung – **to resume** wieder aufnehmen – **amazing** erstaunlich – **to join** zusammenkommen, -treffen – **satisfied** zufrieden(stellend) – **achievements** Erreichtes – **rewarding** lohnend – **enquiry** Anfrage – **enthusiastic** begeistert – **wear control system** Verschleiss-Überwachungssystem – **impressed** beeindruckt – **base plate drawing** Zeichnung der Grundplatte – **to expect** erwarten – **to keep up to date** auf dem laufenden halten – **prospective** zu erwartend, möglich – **sufficient** ausreichend – **routine mail** routinemässige Post – **out of the ordinary** Ungewöhnliches – **to turn up** vorkommen, passieren – **urgent** dringend – **slide** Diapositiv – **educational film** Lehrfilm – **available** erhältlich – **passengers are requested** Passagiere werden gebeten

# *Meeting the experts*

## Meeting a customer for the first time

**Good morning, Mr. Kimball. I'm (Mr.) Gysin.**
Guten Morgen, Mr. Kimball. Ich bin (Herr, Mr.) Gysin

**You've been here before, I assume?**
Sie waren schon einmal hier, nehme ich an?

**Yes, once (twice, three times).**
Ja, einmal (zweimal, dreimal).

**Last time I came here I met Mr. Ammann.**
Als ich das letztemal hier war, war ich mit Herrn Ammann zusammen.

**If you agree, Mr. Gysin, I'd like to meet Mr. Ammann this evening, too.**
Wenn es Ihnen recht ist, Herr Gysin, würde ich heute abend auch gern mit Herrn Ammann sprechen.

**Last time, I was here for four weeks.**
Letztes Mal war ich vier Wochen lang hier.

**How was your flight across the Atlantic?**
Wie war Ihr Flug über den Atlantik?

**Marvellous. It was a unique, quiet flight.**
Wunderbar. Es war ein einmaliger, ruhiger Flug.

**Would you like a cup of coffee or something to eat?**
Möchten Sie (gern) eine Tasse Kaffee oder etwas zu essen?

**I could do with a cup of coffee.**
Eine Tasse Kaffee würde mir guttun (oder: könnte ich gebrauchen).

**Would you like me to arrange a meeting with Mr. Keller?**
Soll ich eine Besprechung mit Herrn Keller organisieren?

**To start with, I'd like to talk to Mr. Ehrensperger, the chief test engineer.**
Zuerst mal würde ich gern mit Herrn Ehrensperger sprechen, dem Prüffeldleiter.

**Is Mr. Zellweger also available?**
Ist Herr Zellweger auch da (oder: zu sprechen)?

**I've had several talks with him on the telephone, and this is a good opportunity to meet him personally.**
Ich habe mit ihm schon mehrmals am Telefon gesprochen, und dies ist eine gute Gelegenheit, ihn persönlich zu sprechen.

**Mr. Zellweger's office is next door.**
Herrn Zellwegers Büro ist (gleich) nebenan.

**I'll see whether he's in.**
Ich werde mal nachsehen, ob er drin ist.

**How long do you intend staying with us?**
Wie lange meinen Sie, bei uns zu bleiben?

**I'll be here for five days this time.**
Diesmal werde ich fünf Tage (lang) bleiben.

**I must be back in New York by Monday at the latest.**
Spätestens am Montag muss ich in New York zurück sein.

**Next time, I hope, you'll bring Mr. Brasnet with you.**
Ich hoffe, Sie bringen das nächstemal Mr. Brasnet mit.

**'I just met the most incredible aluminium siding salesman ...'**

# Enquiring about production progress

**Mr. Gysin, can you tell me, how has production progressed since my last visit?**
Herr Gysin, können Sie mir sagen, wie Sie mit der Fertigung seit meinem letzten Besuch vorangekommen sind?

**All control connections have been completed, and the controllers are now being assembled.**
Alle Steueranschlüsse sind fertig, und die Steuergeräte werden nun zusammengebaut.

**Will you be able to meet the planned delivery dates?**
Werden Sie die geplanten Ablieferungstermine einhalten können?

**As to this type of equipment, yes.**
Was diese Art von Geräten betrifft, ja.

**And we assume the tests will be successfully completed.**
Und (dabei) nehmen wir an, dass die Prüfungen erfolgreich verlaufen.

**Before I go, could you show me the first test results again.**
Können Sie mir noch einmal die ersten Prüfungsergebnisse zeigen, bevor ich gehe.

**Yes, of course. In fact, I can let you have a copy if you wish.**
Aber selbstverständlich. Ich kann Ihnen auch eine Kopie mitgeben, wenn Sie möchten.

**I've arranged a meeting at 11 o'clock.**
Ich habe eine Besprechung für 11 Uhr arrangiert.

**Good, that gives us time to have a quick look at the interface problem.**
Gut, dann haben wir (noch) Zeit, schnell mal das Problem mit der Schnittstelle zu beleuchten.

**Would you like to visit the control section today?**
Möchten Sie heute noch die Abteilung für Steuerungen besichtigen?

**I don't think I'll be able to fit it in today, but perhaps during my next visit.**
Das bringe ich wahrscheinlich heute nicht mehr unter, aber vielleicht bei meinem nächsten Besuch.

# At the departmental meeting

**Let's arrange a meeting with the control section.**
Wir werden eine Besprechung mit der Steuerungs-Gruppe vorbereiten.

**It won't take very long.**
Es wird nicht sehr lange dauern.

**Mr. Wilkinson, the managing director, wants to see us at ten fifteen.**
Mr. Wilkinson, der Direktor, möchte uns um zehn Uhr fünfzehn sprechen.

**We should draw up a capsule appraisal.**
Wir sollten eine kurz(ab)gefasste Abschätzung der Lage entwerfen.

**First of all we should tackle the display problem.**
Zuerst sollten wir mal mit dem Anzeige-Problem fertig werden.

**This is only a rough-and-ready guide to our display system.**
Dies ist nur eine (sehr) grobe Vorstellung von unserem Anzeigesystem.

**Mr. Burger has put forward a number of good ideas which make good sense.**
Mr. Burger hat eine Reihe guter Vorschläge vorgelegt, die sich als (durchaus) sinnvoll erweisen.

**The most recent modifications on these diagrams are marked in red.**
Die letzten Änderungen auf diesen Plänen sind in Rot eingetragen.

**The progress report for the Trowbridge scheme will be issued tomorrow.**
Der Montagebericht für die Anlage Trowbridge wird morgen herausgegeben.

**Can we now revert to the components to be supplied by our sub-contractors.**
Können wir nun auf die Teile zurückkommen, die von unseren Unterlieferanten zugeliefert werden.

**We would be very competitive if it weren't for those ECL circuits as per specification.**
Wir könnten sehr konkurrenzfähig sein, wenn da nicht, laut Spezifikation, die ECL-Schaltkreise wären.

**There's still something else that bothers me.**
Da ist noch etwas anderes, was mir Kummer bereitet.

**We've got to order control panels ourselves.**
Wir müssen die Steuertafeln selber bestellen.

**Now to something more promising.**
Nun zu etwas Erfreulicherem.

**We shall be awarded, so Mr. White told me, the whole Carlisle order.**
Wir werden den kompletten Carlisle-Auftrag zuerkannt bekommen,
das erzählte mit Mr. White.

**Now to the last item but one of this meeting.**
Wir kommen jetzt zum zweitletzten Punkt dieser Besprechung.

**I'll tell you first off the record.**
Ich möchte Ihnen erstmal (dieses) am Rande erzählen.

**We shouldn't interfere too much with Mr. Money's work.**
Wir sollten uns nicht zu sehr in die Arbeit von Mr. Money einmischen.

**MinProg activity**

**Latest MinProg achievement. An artist's impression of the phantastic
flowers, intended for use in large-room offices, as developed at their
laboratories in Pulham Down. The Ministry of Progress is always
ahead of time, so a spokesman told New Scientist recently.**

# Telephone connection to America

Starting a telephone call one should say, after greeting, the name, followed by the company and the place.

**Good morning. This is (Mr.) Leue, Metrohm Company, speaking from Erlenbach.**
If you have to repeat, go on like this:

**My name is Leue, Metrohm Company ...**

**Good afternoon. My name ist Leue, Metrohm Company, speaking from Erlenbach.**
Guten Tag. Mein Name ist Leue, Firma Metrohm, in Erlenbach.

**I'd like to speak (or: talk) to Mr. Hanson.**
Ich möchte gern mit Herrn Hanson sprechen.

**Just a moment. I'll see if Mr. Hanson is available (or: free).**
Einen Moment, ich werde mal nachsehen, ob Herr Hanson da ist (oder: frei ist).

**Mr. Hanson, a Mr. Leue is on the line for you.**
Herr Hanson, ein Herr Leue möchte Sie am Telefon sprechen.

**Put him through, please.**
Stellen Sie ihn bitte durch.

**Mr. Leue, I'm putting you through now.**
Herr Leue, ich stelle Sie jetzt durch.

**Hold the line please.**
Bleiben Sie bitte am Apparat.

**Please give my regards to Mr. Wilson (or: remember me to Mr. Wilson; or: extend my greetings to Mr. Wilson).**
Bestellen Sie bitte Herrn Wilson einen schönen Gruss.

**Goodbye, Mr. Leue, thanks for calling.**
Auf Wiedersehen, Herr Leue, vielen Dank für den Anruf.

# Sorry, but I didn't catch your name

**Good afternoon. My name is Randall, Primotec Company, speaking from Taunton.**
Guten Tag. Mein Name ist Randall, Firma Primotec, in Taunton.

**I'd like to speak to Mr. Brunner.**
Ich möchte gern mit Herrn Brunner sprechen.

**I'm sorry, but I didn't catch your name.**
Es tut mir leid, aber ich habe Ihren Namen nicht verstanden.

**I'll spell it out.**
Ich werde (meinen Namen) buchstabieren.

**Robert – Andrew – Nellie – David – Andrew – Lucy – Lucy.**
Rosa – Anna – Niklaus – Daniel – Anna – Leopold – Leopold.

**Just a moment. I'll see if Mr. Brunner is free.**
Einen Moment mal. Ich werde nachsehen, ob Herr Brunner frei ist.

**Mr. Randall, I'm putting you through now.**
Herr Randall, ich stelle Sie jetzt durch.

**Would you speak up please.**
Würden Sie bitte etwas lauter sprechen.

**The connection is very bad. I can hardly understand you.**
Die Verbindung ist sehr schlecht. Ich kann Sie kaum verstehen.

**Hang up please. I'll ring you again immediately. I hope the connection will be better then.**
Legen Sie bitte auf. Ich rufe gleich wieder an. Ich hoffe, dass die Verbindung dann besser ist.

**What time shall I ring you tomorrow?**
Um welche Zeit soll ich Sie morgen anrufen?

**Please give me Mr. Imhof again.**
Geben Sie mir bitte noch einmal Herrn Imhof.

**Am I speaking to the chief test engineer?**
Spreche ich mit dem Prüffeld-Leiter?

**Would you please repeat that.**  **Goodbye Mr. Brunner.**
Würden Sie das bitte wiederholen.  Auf Wiedersehen Herr Brunner.

# Phonetic alphabets

## Buchstabiertabelle

|   | Switzerland | Great Britain | United States | Germany |
| --- | --- | --- | --- | --- |
| **A** | Anna | Andrew | Abel | Anton |
| **B** | Bertha | Benjamin | Baker | Berta |
| **C** | Cäsar | Charlie | Charlie | Cäsar |
| **D** | Daniel | David | Dog | Dora |
| **E** | Emil | Edward | Easy | Emil |
| **F** | Friedrich | Frederick | Fox | Friedrich |
| **G** | Gustav | George | George | Gustav |
| **H** | Heinrich | Harry | How | Heinrich |
| **I** | Ida | Isaac | Item | Ida |
| **J** | Jakob | Jack | Jig | Julius |
| **K** | Kaiser | King | King | Kaufmann |
| **L** | Leopold | Lucy | Love | Ludwig |
| **M** | Marie | Mary | Mike | Martha |
| **N** | Niklaus | Nellie | Nan | Nordpol |
| **O** | Otto | Oliver | Oboe | Otto |
| **P** | Peter | Peter | Peter | Paula |
| **Q** | Quelle | Quennie | Queen | Quelle |
| **R** | Rosa | Robert | Roger | Richard |
| **S** | Sophie | Sugar | Sugar | Samuel |
| **T** | Theodor | Tommy | Tare | Theodor |
| **U** | Ulrich | Uncle | Uncle | Ulrich |
| **V** | Viktor | Victor | Victor | Viktor |
| **W** | Wilhelm | William | William | Wilhelm |
| **X** | Xaver | Xmas | X | Xantippe |
| **Y** | Yverdon | Yellow | Yoke | Ypsilon |
| **Z** | Zürich | Zebra | Zebra | Zacharias |

# Choosing the correct meeting times

Correct use of prepositions requires a sound knowledge of the English language. Some examples are given below to help the student from making typical errors. Emphasis is put on prepositions 'on', 'at', 'by' and 'in'.

**When will we be meeting again?**
Alternative: When do we meet again?
The word 'do' suggests the date is already fixed. Further alternative: When should (or: can) we meet again?
Wann findet unsere nächste Sitzung (oder: Tagung) statt?

**We'll hold our next meeting on 10th June, at 10.30 a. m. in our office.**
Unsere nächste Sitzung findet am 10. Juni, um 10 Uhr dreissig, in unserem Hause statt.

**Who will be organizing the meeting?**
Wer wird die Sitzung organisieren?

**The meeting will be organized by Mrs. Westwood.**
Die Sitzung wird von Mrs. Westwood organisiert.

**I'll let you know the details by Tuesday at the latest.**
Alles Nähere teile ich Ihnen bis spätestens Dienstag mit.

**By what date can you let me know whether you will be attending our meeting?**
Bis wann können Sie mir mitteilen, ob Sie an unserer Sitzung teilnehmen werden?

**I'll be in New York for about ten days, but hope to be back by then.**
Ich werde etwa zehn Tage in New York sein, hoffe aber bis dahin zurück zu sein.

**Couldn't we postpone the meeting until 20th June?**
Können wir nicht die Sitzung bis zum 20. Juni verschieben?

**That would be all right with (or: by) me.**
Alternative: That's alright by me.
Von mir aus ist das in Ordnung. (Oder: Das passt mir gut.)

**Why not bring the meeting forward to Monday?**
Warum können wir die Sitzung nicht auf Montag vorverlegen?

**There you are. I've been looking for you all over the place.**
Da sind Sie ja. Ich habe Sie überall gesucht.

**Mr. Keller just told me he couldn't attend the meeting that early.**
Herr Keller sagte mir gerade, dass er an der Sitzung nicht so früh teilnehmen könnte.

**You'll receive the minutes of meeting by Wednesday at the latest.**
Here 'by' means possibly earlier but not later than.
Die Besprechungsniederschrift erhalten Sie spätestens bis am Montag.

'Ms. Ryan, send me in a scapegoat.' (Scapegoat: Sündenbock)

# 'Speaking about' and 'writing on' an engineering scheme

Mistakes are frequently made when applying ‹on› and ‹about›. The following examples demonstrate correct usage of these two very common and tricky prepositions.

**We were talking about the control scheme.**
Wir sprachen über das Steuerungssystem.

**He gave a talk on the control scheme. (Talk here: lecture.)**
Er hielt eine (kleine) Vorlesung über das Steuerungssystem.

**He wrote an article on the control scheme.**
Er schrieb einen Artikel über das Steuerungssystem.

**He wrote about the control scheme.**
Er schrieb über das Steuerungssystem.

**We talked about the Sulzer Z engines.**
Wir sprachen über die Sulzer-Z-(Diesel-)Motoren.

**The engineer wrote an article on the Sulzer Z engines.**
Der Ingenieur schrieb einen Artikel über die Sulzer-Z-(Diesel-)Motoren.

**The engineer wrote about the Sulzer Z engines.**
Der Ingenieur schrieb über (oder: beschrieb) die Sulzer-Z-(Diesel-)Motoren.

*Other ways of expressing ‹talk about›*
We discussed the control scheme in detail.
We described the control scheme.
We outlined the control scheme.
We went over (the problems on) the control scheme.
We talked over (the problems concerning) the control scheme.

Using other tenses we can also say:
We were discussing...
We had discussions about...
We were having discussions about...

# Customer comments on diagrams

**We do not approve of the use of these semi-enclosed electric motors.**
Wir gestatten es nicht, diese halbgeschlossenen Elektromotoren
(geschützt gegen zufällige Berührung) zu verwenden.

**We recommend the use of totally enclosed motors in damp rooms.**
Wir empfehlen, vollkommen geschlossene Motoren in feuchten
Räumen zu verwenden.

**We should like to ask you to substitute sirens for bells.**
Wir möchten Sie bitten, Sirenen an Stelle von Glocken vorzusehen.

**As a matter of principle we do not accept a higher service voltage for
this installation**
Wir akzeptieren für diese Anlage grundsätzlich keine höhere Betriebs-
spannung.

**The characteristics of chargers and batteries are not matched.**
Die Charakteristiken der Ladegeräte und Batterien stimmen nicht
überein.

**The tank gauging system must operate at (or: on) 24 V d. c.**
Die Tankmessanlage ist mit 24 V Gleichspannung zu betreiben.

**The specification calls for relays of Brown & Sharpe manufacture.**
Die Spezifikation schreibt Relais der Firma Brown & Sharpe vor.

**We note that all electric motors for essential services are fitted with
Class B insulation.**
Wir nehmen zur Kenntnis, dass alle Elektromotoren für wichtige
(wörtlich: wesentliche) Betriebe in Isolationsklasse B ausgeführt sind.

**We should prefer an electronically operated remote metering system.**
Wir würden einem elektronisch betriebenen Fernmessgerät den Vorzug geben.

**Intrinsically safe equipment and associated circuits are physically
separated from other circuits.**
Eigensichere Anlagen und die dazugehörigen Stromkreise sind von
den anderen Stromkreisen physikalisch getrennt.

**We regret to have to inform you that the operation of the phase comparison device cannot be derived from the drawing submitted.**
Wir bedauern, Ihnen mitteilen zu müssen, dass die Arbeitsweise des Phasenvergleichsgerätes von den übersandten Zeichnungen nicht abgeleitet werden kann.

**Our tools are exposed to severe tests.**

# Repairing faulty equipment

**First of all, we'll have to remove the top cover.**
Als erstes müssen wir die obere Abdeckung (oder: den Deckel) entfernen (oder: abnehmen).

**Then we must remove the rear plate.**
Dann müssen wir die hintere Platte abnehmen (oder: entfernen).

**Now I'm going to draw out the printed-circuit board.**
Jetzt werde ich die Leiterplatte herausziehen.

**I'm disconnecting the feeder cable now.**
Nun werde ich das Speisekabel entfernen (oder: abklemmen, lösen).

**Both current transformers must also be removed.**
Beide Stromwandler müssen auch ausgebaut (oder: entfernt, herausgenommen) werden.

**The terminal block is to be screwed off.**
Der Klemmblock ist abzuschrauben.

**The second diode is defective.**
Die zweite Diode ist defekt.

**Hand me the little screw-driver, please.**
Reich mir bitte den kleinen Schraubenzieher.

**I'm removing the diode now.**
Ich nehme jetzt die Diode heraus.

**We needn't strip down the whole device.**
Wir brauchen nicht das ganze Gerät auseinanderzunehmen.

**Would you please check the diode.**
Würdest du bitte die Diode prüfen.

**Is there a spare diode?**
Ist da (noch) eine Reservediode?

**I'm afraid there's no spare diode.**
Ich fürchte, wir haben keine Reservediode.

**Don't worry, Metroniks carry this type of diode on stock.**
Keine Bange, Metroniks hat diesen Diodentyp auf Lager.

# *Common phrases as used abroad*

## At the airport

**Where is the airline's town office?**
Wo ist das Stadtbüro dieser Fluggesellschaft?

**At the corner.**
An der Ecke.

**How much is a return flight to Zurich?**
Was kostet ein Hin- und Rückflug nach Zürich?

**What is the free baggage allowance?**
Wieviel Gepäck ist frei?

**45 pounds (20 kg) are free.**
20 kg sind frei.

**What is the excessive baggage rate per pound?**
Was ist für ein weiteres Kilo zu zahlen?

**When does the next plane leave?**
Um wieviel Uhr startet das nächste Flugzeug?

**10.20 a. m. Would you like a time-table?**
10.20 Uhr. Möchten Sie einen Flugplan?

**When must I be at the airport?**
Wann muss ich auf dem Flughafen sein?

**Three quarters of an hour before your plane leaves.**
Eine Dreiviertelstunde vor Abflug.

**How do I get to the airport?**
Wie komme ich zum Flughafen?

**The best thing to do is to take the airport bus.**
Am besten nehmen Sie den Zubringerbus.

## At the station

**Is there a train to Trowbridge in the morning?**
Fährt vormittags ein Zug nach Trowbridge?

**Yes, at 11.25 a.m.**
Ja, 11.25 Uhr.

**Do I have to change?**
Muss ich umsteigen?

**No, you take the through-coach.**
Nein, Sie nehmen den Kurswagen.

**Is there a connection to Schenectady?**
Habe ich Anschluss nach Schenectady?

**No, but you have only half an hour's wait.**
Nein, Sie haben nur eine halbe Stunde Aufenthalt.

**Is there a restaurant-car on the train?**
Hat der Zug einen Speisewagen?

**No, this train has only a refreshment-car.**
Nein, dieser Zug führt nur einen Büffetwagen.

**When do I arrive?**
Wann komme ich an?

**At 2 p.m.**
Um 14 Uhr.

**One first class ticket, and a reserved seat, a window seat, please.**
Bitte eine Fahrkarte 1. Klasse und eine Platzkarte, Fensterplatz.

**Which platform does this train leave from?**
Von welchem Bahnsteig fährt der Zug ab?

**Platform 14.**
Bahnsteig 14.

**Thank you very much.**
Vielen Dank.

## At the customs

**Have you anything to declare?**
Haben Sie etwas zu verzollen?

**No, I have nothing to declare.**
Nein, ich habe nichts zu verzollen.

**I only have a bottle of spirits.**
Ich habe nur eine Flasche Alkohol.

**Open this suitcase.**
Öffnen Sie diese Tasche.

**I have lost my keys.**
Ich habe meine Schlüssel verloren.

**I cannot open my case.**
Ich kann diese Tasche nicht öffnen.

**What is in the box?**
Was ist in der Kiste?

**There is nothing but ...**
Da sind nur ...

**All that is for my personal use.**
Alles ist für den persönlichen Gebrauch.

**This is already marked.**
Dies ist schon bezeichnet.

**Help me to close this case.**
Helfen Sie mir, diese Tasche zu schliessen.

# At the hotel

**I wrote to you three weeks ago.**
Ich habe Ihnen vor drei Wochen geschrieben.

**Did you not get my letter?**
Haben Sie meinen Brief nicht erhalten.

**I asked for a second-floor room.**
Ich bat um ein Zimmer im zweiten Stock.

**Can I have a room for one night?**
Kann ich ein Zimmer für eine Nacht haben?

**I intend to stay a week at least.**
Ich beabsichtige, mindestens eine Woche zu bleiben.

**I am only staying for two or three days.**
Ich bleibe nur zwei oder drei Tage.

**Have you got a room on the first floor?**
Haben Sie ein Zimmer im ersten Stock?

**What is the price of a room per night?**
Was kostet das Zimmer pro Nacht?

**I want a bedroom with breakfast only.**
Ich möchte ein Zimmer nur mit Frühstück.

**I want a quiet room.**
Ich möchte ein ruhiges Zimmer.

**I do not want a room looking on to the street.**
Ich möchte kein Zimmer mit Blick auf die Strasse.

**Very well, I shall take this room.**
In Ordnung, ich nehme dieses Zimmer.

# In a restaurant

**Can we lunch (or: dine) here?**
Können wir hier essen?

**There are four of us.**
Wir sind vier Personen.

**Do you have a menu card?**
Haben Sie eine Speisekarte?

**We do not want a complete meal.**
Wir möchten kein komplettes Mahl.

**We only want a snack.**
Wir möchten nur einen Imbiss.

**We are in a hurry.**
Wir sind in Eile.

**We want to dine at once.**
Wir möchten sofort essen.

**We shall come back at (or: by) one o'clock.**
Wir kommen um (oder: bis) ein Uhr zurück.

**I should like to wash my hands before the meal.**
Ich möchte meine Hände vor dem Essen waschen.

**The toilet is this way, Sir.**
Die Toilette finden Sie dort (entlang).

**What would you like to drink?**
Was möchten Sie trinken?

**Bring us the wine-list, please.**
Bringen Sie uns bitte die Weinliste.

**I am very thirsty. I want something refreshing.**
Ich habe grossen Durst. Ich möchte etwas Erfrischendes.

**Would you like some coffee?**
Möchten Sie gern Kaffee?

**The bill, please.**
Ich möchte zahlen (oder: die Rechnung bitte).

**Is the service (or: the cover charge) included?**
Ist Bedienung inbegriffen?

**That's just right.**
Das stimmt so.

**There is a mistake in the bill.**
Da ist ein Fehler in der Rechnung.

**I made a mistake.**
Ich habe einen Fehler gemacht.

**Please don't mention it.**
Nicht der Rede wert

**My Sahara-Crossing Slimming holidays are a huge success. Package tours without food, drink or accommodation are available from Fr. 3000 per week.**

# Public notices

**It is forbidden to ...**
Es ist verboten ...

**No ... Trespassers will be prosecuted.**
Es ist verboten ... Verstösse werden geahndet.

**Spitting is forbidden.**
Spucken verboten.

**Stick no bills.**
Plakate ankleben verboten.

**Tourists are requested to ...**
Touristen werden gebeten ...

**Keep to the left (right).**
Links (rechts) halten.

**Beware of the dog.**
Vorsicht bissiger Hund.

**Please wipe your feet.**
Bitte Füsse abtreten.

**Vacant. Engaged.**
Frei. Besetzt.

**Knock. Ring.**
Klopfen. Läuten (oder: Klingeln).

**Wet paint.**
Frisch gestrichen.

**No entry.**
Eintritt verboten.

**Closed for holidays.**
Wegen Ferien geschlossen.

# Expressing praise

**You've done a fine job.**
Das haben sie gut gemacht.

**I'm very grateful, you've done all you can.**
Ich bin Ihnen sehr dankbar, Sie haben getan, was Sie konnten.

**You've put it very nicely.**
Das haben Sie nett gesagt.

**I can't but say a word of praise to Mr. Russell.**
Ich kann nicht umhin, Mr. Russell besonders zu erwähnen (oder: zu loben).

**It's very wise of you.**
Das ist sehr klug von Ihnen.

**I declare you've done fine work.**
Ich muss schon sagen, Sie haben gute Arbeit geleistet.

**You can take pride in that design.**
Auf diese Konstruktion können Sie stolz sein.

**I'm glad you've been looking after that business.**
Ich bin froh, dass Sie sich um diese Angelegenheit gekümmert haben.

**You've got a good command of the German language.**
Sie beherrschen die deutsche Sprache ausgezeichnet.

**I'm happy you'll assist me with this job.**
Ich bin froh, dass Sie mich bei dieser Arbeit unterstützen wollen.

**I've never seen a layout like this before.**
Ich habe bisher noch keinen Entwurf wie diesen gesehen.

**I'm sure you'll manage it somehow.**
Ich bin sicher, dass Sie es (schon) irgendwie meistern werden.

## Calming down people

**Never mind, it's natural error (or: it can happen to anybody).**
Macht nichts, (das) kann jedem passieren.

**Think nothing of it.**
Macht nichts (oder: mach dir keine Gedanken).

**Put yourself at ease, we have not lost yet.**
Beruhige dich, noch haben wir nicht verloren.

**Unbend a little, it isn't his fault either.**
Sei nachgiebig, seine Schuld ist es auch nicht (wörtlich: lockere deine starre Haltung).

**Take it easy, we still have got time to start from scratch.**
Nimm es leicht, noch haben wir Zeit, ganz von vorn zu beginnen.

**No use moaning about it, we can't do any more in this case.**
Es hat keinen Zweck herumzujammern, wir können in diesem Fall nichts mehr tun.

**Calm down, Eugene, I really didn't mean it.**
Beruhige dich, Eugene, ich habe es wirklich nicht so gemeint.

**I wouldn't dream of blaming you for that, we should have seen it, too.**
Ich denke nicht daran, dir die Schuld zu geben, wir hätten es auch sehen sollen.

**I'm afraid you must have thought me extremely rude but I was very tired. I haven't had any sleep for two days.**
Ich fürchte, Sie haben mich für sehr grob gehalten, aber ich war sehr müde. Ich hatte zwei Tage keinen Schlaf gehabt.

**Don't take it amiss, we all make mistakes in a situation like this.**
Nimm es (mir) nicht übel, wir alle machen Fehler in solchen Situationen.

# Agreement

**I agree with you entirely.**
Ich stimme Ihnen vollkommen zu.

**I fully agree here.**
Hier stimme ich Ihnen voll zu.

**It looks all right to me.**
Es scheint mir in Ordnung zu sein.

**You can certainly say that again.**
Das kann man wohl sagen.

**It's quite right with us.**
Bei uns ist das in Ordnung (auch: von uns aus ist ...).

**That's just what I was going to say.**
Genau das wollte ich sagen.

**You are absolutely right.**
Sie haben vollkommen recht.

**That will do fine.**
Das trifft sich gut (oder: das geht gut).

**Quite so. Sounds good.**
Das stimmt. Lässt sich hören.

**I quite see that.**
Ich stimme hier zu.

**I think so. Right you are.**
Ich denke ja. Sie haben recht.

**I see the point.**
Ich sehe (schon) was Sie meinen.

**He thought so as well.**
Er dachte das gleiche.

**I'm sure you know the problem better than I do.**
Ich bin sicher, dass Sie sich hier besser auskennen als ich.

**We'll agree on this subject soon.**
Bei diesem Thema werden wir uns schnell einig.

# Disagreement

**I can't possibly agree to your proposal.**
Ihrem Vorschlag kann ich unmöglich zustimmen.

**Things are different with us.**
Bei uns ist das anders.

**It's not quite so simple as you think.**
Es ist nicht ganz so einfach wie Sie denken.

**It's not as simple as that.**
So einfach ist das auch (wieder) nicht.

**I won't consider that.**
Das kommt nicht in Frage.

**I'm not going to argue about something so absurd.**
Ich werde mich über so alberne Sachen nicht streiten.

**I don't like the sound of it.**
Das gefällt mir gar nicht.

**What's the use of going on if you don't agree on this item.**
Was hat es für einen Zweck weiterzumachen, wenn Sie in diesem Punkt nicht zustimmen.

**I felt all along that something was wrong here.**
Ich dachte schon die ganze Zeit, dass hier etwas nicht stimmen könnte.

**That's not at all clear to me.**
Das ist mir durchaus nicht klar.

**One really oughtn't to say that.**
Das sollte man wirklich nicht sagen.

**You can never hope to achieve the same results again.**
Diese Ergebnisse werden Sie nicht noch einmal erreichen.

**I don't think we should leave it as it is.**
Ich meine, wir sollten es nicht so lassen.

**I see absolutely no reason why I shouldn't go on like this.**
Ich sehe überhaupt keinen Grund, warum ich nicht so weitermachen sollte.

**I've told him again and again that we can't go on like this.**
Ich habe ihm immer wieder gesagt, dass wir so nicht weitermachen
können.

**Nothing doing, we must turn down your request.**
Nichts zu machen, wir müssen Ihr Gesuch ablehnen.

**Catch me ever admitting that.**
Da können Sie lange warten bis ich das zugebe.

**You don't say. Don't side-track.**
Was Sie da nicht sagen. Weichen Sie nicht vom Thema ab.

**What's that to you?**
Was geht Sie das an?

**That's not your business.**
Das geht Sie gar nichts an.

**That's too much.**
Das geht (doch) zu weit.

**Don't drag me into that affair.**
Lassen Sie mich aus dem Spiel.

**It's all over and done with.**
Es ist alles aus (und vorbei).

**Don't think about it any longer.**
Schlagen Sie es sich aus dem Sinn.

**We have not seen the last of it yet.**
Es ist noch nicht aller Tage Abend.

**We are arguing at cross-purposes.**
Wir reden aneinander vorbei.

**You'd better leave that alone.**
Sie sollten lieber die Finger davon lassen.

# Thinking over, thinking up

**We must think it over again.**
Wir müssen es uns nochmal überlegen.

**I was just thinking something out.**
Ich dachte gerade über etwas nach (im Sinne von: zu Ende denken).

**Let's think something up.**
Denken wir uns etwas aus.

**Who would have thought of that.**
Wer hätte das gedacht.

**When you come to think of it.**
Wenn man es sich richtig überlegt.

**Think about it.**
Denken Sie doch mal nach.

**They half thought of starting again.**
Sie dachten fast daran, noch einmal anzufangen.

**They think a lot of Mr. Hansen.**
Sie halten viel von Herrn Hansen.

**When I heard the truth about Mr. Keller, I thought no end of him.**
Als ich die Wahrheit über Herrn Keller erfuhr, bekam ich eine ungemein hohe Meinung von ihm.

**We don't think much of this gadget.**
Wir halten nicht viel von diesem Gerät («Dingsda»).

**I think it's advisable to leave it as it is.**
Ich halte es für ratsam, es so zu lassen, wie es ist.

**All thinking men would do it this way.**
Jeder vernünftig Denkende würde es so machen.

**I must do some hard thinking.**
Ich muss mal scharf nachdenken.

**Don't you think you ought to send for an expert.**
Meinst du nicht, dass du einen Fachmann holen solltest.

**Though on second thoughts, I wonder whether ...**
Doch bei nochmaliger Überlegung frage ich mich, ob ...

**It was enough to make me think.**
Es war genug, meine Aufmerksamkeit zu erregen.

**'The S.O.B.s are getting too damn' confident.'**

# General questions concerning work

**I want to (get to) know if we can start now.**
Ich möchte wissen, ob wir nun beginnen können.

**What do you mean by that?**
Was soll das heissen?

**Do you happen to know the XY formula?**
Kennen Sie zufällig die XY-Formel.

**Can you make yourself understood more clearly.**
Können Sie sich (nicht) klarer ausdrücken.

**I didn't mean to interrupt, but we must revert to that point again.**
Ich wollte nicht unterbrechen, aber wir müssen auf diesen Punkt noch einmal zurückkommen.

**Whatever for? What's on at XY?**
Wofür bloss? Was ist bei XY los?

**What makes you say that?**
Warum glauben Sie das?

**What does the safety paragraph read?**
Wie lautet der Sicherheitsabschnitt?

**What's this got to do with our device?**
Was hat das mit unserem Gerät zu tun?

**Haven't you wanted to modify the controller?**
Wollten Sie nicht das Steuergerät abändern?

**Are you mechanically minded?**
Sind Sie handwerklich begabt?

**Has it proved of any value?**
Hat es sich (überhaupt) als nützlich (oder: von Wert) erwiesen?

**Can you give me a ring tomorrow?**
Können Sie mich morgen anrufen?

**Can you make yourself understood more clearly.**
Können Sie sich nicht klarer ausdrücken?

**This doesn't work.**
Das geht (so) nicht.

**On the contrary.**
Im Gegenteil.

**It's fine (bad) weather.**
Es ist gutes (schlechtes) Wetter.

**I think so.**
Ich glaube (schon).

**I don't think so.**
Das glaube ich nicht.

**Listen. Look.**
Hören Sie. Sehen Sie.

**Very well. That's all.**
Sehr gut. Das ist alles.

**Whose turn is it?**
Wer ist an der Reihe?

**It is not my fault.**
Es ist nicht meine Schuld.

**I am looking for ...**
Ich suche (nach) ...

**It is very annoying.**
Das stört mich sehr.

**It has nothing to do with me.**
Dafür kann ich (wirklich) nichts.

**Where should one apply?**
Wohin sollte man sich wenden.

**What does that mean?**
Was bedeutet das?

**Take it easy.**
Nehmen Sie es leicht.

**You are requested to ..**
Sie werden gebeten ...

**What are you driving at?**
Worauf wollen Sie hinaus?

## Some useful sayings

**though on second thoughts**
doch bei nochmaliger Überlegung

**to be perfectly candid**
um ganz ehrlich zu sein

**to look after a person**
sich um jemanden zu kümmern

**just stop and think**
überlegen Sie einmal

**to make out a case for**
ein Argument für … vorbringen

**the problems raised**
die sich daraus ergebenden Probleme

**to be under moral obligation**
eine moralische Verpflichtung haben

**to fill a practical purpose**
einen praktischen Sinn erfüllen

**at the expense of**
auf Kosten … (Genitiv)

**if it comes to that**
wenn es darauf hinausläuft

**to go about doing**
sich daran machen

**after all eventually**
schliesslich, mit der Zeit

**to take a bashing (fam.)**
einen schweren Schlag bekommen

**to concern oneself with**
sich befassen mit

**to lend one's name for**
seinen Namen hergeben für

**just as you care to take it**
wie man es nimmt

**be that as it may**
es mag sein, wie es will

**to chiv(v)y a person**
jemanden herumjagen

**in a nutshell**
in aller Kürze

**to have a good chat**
sich ausgiebig unterhalten

**in this day and age**
heutzutage

**at your discretion**
nach Ihrem Ermessen

**to take out of circulation**
aus dem Verkehr ziehen

**to keep tabs on a person**
jemanden ständig beobachten

**to take into consideration**
in Betracht ziehen

**he has got something in his point**
er hat schon recht mit seiner Behauptung

**to think it pretty significant**
es für ziemlich wichtig halten


## Polite expressions


**Oh yes. Oh no.**
Aber ja. Aber nein.

**I beg your pardon.**
Ich bitte um Entschuldigung.

**Excuse me.**
Entschuldigen Sie mich.

**That's all right.**
Das ist in Ordnung.

**Don't mention it.**
Nicht der Rede wert.

**Please sit down.**
Setzen Sie sich bitte.

**Can I help you?**
Kann ich Ihnen behilflich sein?

**Help yourself!**
Bedienen Sie sich (doch)!

**How are you?**
Wie geht es Ihnen?

**I am very well, thank you.**
Danke, mir geht es (sehr) gut.

**I'm fine, thanks.**
Mir geht es gut, danke.

**How do you do (delighted to meet you).**
Etwa: Guten Tag (freue mich, Sie kennen zu lernen).

**Allow me.**
Gestatten Sie.

**You are very kind.**
Sie sind sehr nett.

**Am I disturbing you?**
Störe ich Sie?

**I am terribly sorry.**
Das tut mir sehr leid.

**Don't worry.**
Machen Sie sich keine Sorge.

**Carry on.**
Machen Sie (nur) weiter.

**Your good health.**
Auf Ihre Gesundheit.

**Allow me to introduce...**
Gestatten Sie, dass ich Ihnen... vorstelle.

**I am very grateful to you.**
Ich bin Ihnen sehr dankbar.

# Miscellaneous expressions

**Look down there.**
Sehen Sie dort unten (oder: nach unten).

**Look up there**
Sehen Sie dort oben (oder: nach oben).

**That's it.**
Das wär's.

**Quickly. Slowly.**
Schnell. Langsam.

**Look out!**
Geben Sie acht!

**Come in. Go out.**
Herein. Gehen Sie hinaus.

**On this side.**
Auf dieser Seite.

**On the other side.**
Auf der anderen Seite.

**I am an American.**
Ich bin Amerikaner(in).

**What is the matter?**
Was ist los? (oder: was geht da vor?)

**I don't know anything about it.**
Davon weiss ich gar nichts.

**You are right.**
Sie haben recht.

**You are wrong.**
Sie haben nicht recht (oder: Sie sind im Unrecht).

## Hyphen, stroke and slash

**brackets** Klammern; today understood as round brackets; previously the meaning was square brackets.

**broken line** (or: dashed line) gestrichelte Linie; also: to mark with little lines: mit einer gestrichelten Linie versehen.

**dash** Gedankenstrich

**dot-and-dash line** Punkt-Strich-Linie

**dotted line** punktierte Linie; to dot: punktieren

**hyphen** [ˈhaifən] Bindestrich, Trennungszeichen; to hyphen (or: hyphenated) mit Bindestrich versehen.

**parenthesis** [pəˈrenθisis]; Pl.: parentheses: runde Klammern; not frequently used today.

**square brackets** eckige Klammern; to put in square brackets: in eckige Klammern setzen.

**slash** (US) Schrägstrich

**stroke** Schrägstrich, Querstrich; also: diagonal stroke.

## Company administration

**advertising department** Werbeabteilung
**circuit designer** Schaltungsfachmann
**contractor** Unternehmer, Lieferant
**department** Abteilung
**designer** Konstrukteur
**designing department** Konstruktionsabteilung

**division** Geschäftsbereich
**draughtsman** (US: draftsman) Zeichner
**draughtswoman** Zeichnerin
**drawing office** Zeichenbüro
**group** Konzern; US auch: Gruppe
**group leader** (US) Gruppenleiter
**head of department** Abteilungsleiter
**personnel department** Personalabteilung
**public relations officer** Pressereferent
**purchasing department** Einkaufsabteilung
**sales department** Vertriebsabteilung
**sales engineer** (sometimes: applications engineer) Vertriebsingenieur
**section leader** Gruppenleiter
**squad** (US) Gruppe
**sub-contractor** Unterlieferant
**test bay** (or: station, bed, stand) Prüffeld
**training officer** Ausbildungsleiter


## Some useful adverbs of place and time

**here and there** stellenweise
**next door** nebenan
**from behind** von hinten
**at the end** am Ende
**everywhere** überall
**nowhere** nirgends
**at the bottom** ganz unten
**higher up** weiter oben
**anywhere (or: somewhere)** irgendwo
**straight ahead** geradeaus
**halfway between** auf dem halben Wege zwischen
**up to there** bis dahin
**over there** (da) drüben
**up there** da oben
**down there** da unten
**in there** da drin(nen)
**upstairs** oben (im Hause)
**downstairs** unten (im Hause)

**further beside** gleich daneben
**right up** weiter oben
**lower down** weiter unten
**in front** vorn
**in the rear** hinten
**on top** oben (auf)
**from within** von innen
**from without** von aussen
**downward** nach unten
**westward** nach Westen
**to and fro** hin und her
**all over the world** auf der ganzen Welt
**by the end of this month** gegen Ende dieses Monats
**by the mid-eighties** Mitte der achtziger Jahre
**once a month** einmal im Monat
**it was not long before he ...** es dauerte nicht lange bis er ...
**by night** bei Nacht
**last night** gestern abend
**at another time** ein anderes Mal
**(in) these days (or: nowadays)** heutzutage
**one of these days** dieser Tage
**every other day** einen Tag um den anderen, jeden zweiten Tag
**this day (in a) week** heute in acht Tagen
**now and then** hin und wieder
**in between** zwischendurch
**to the last** bis zuletzt
**for good** für immer
**in my time** zu meiner Zeit
**as time goes on** mit der Zeit
**in all his born days** zeit seines Lebens
**at times** zeitweilig, manchmal
**just now** eben noch
**until lately** bis vor kurzem
**the week before last** vorletzte Woche
**all day long** den ganzen Tag
**later on** später
**this time** diesmal
**in due time** rechtzeitig
**ten minutes ahead of time** zehn Minuten vor der Zeit
**inside of two months (US)** innerhalb von zwei Monaten

**for the time being (or: at the present time)** im Augenblick
**the other day** neulich
**on time** pünktlich
**in time to come** künftig, in der Zukunft
**by and by** nach und nach
**at the latest** spätestens
**at the earliest** frühestens
**sooner or later** über kurz oder lang
**formerly** ehemals, früher
**every few weeks** alle paar Wochen
**for a time** eine Zeitlang
**for the first time** zum ersten Male
**for the present** vorläufig
**within memory of man** seit Menschengedenken
**once in a while** gelegentlich, dann und wann
**in the long run** auf die Dauer
**in the meantime** unterdessen, inzwischen
**the very minute (or: the very moment)** gerade in diesem Augenblick
**after a time** nach einiger Zeit
**ever since** von der Zeit an
**right away** sofort
**as late as 1970** noch im Jahre 1970
**as early as Monday** schon Montag
**all of a sudden (or: at once)** ganz plötzlich

## Special adverbs

**above all** vor allem
**for a change** zur Abwechslung
**on this side of forty** unter vierzig Jahren
**by accident** zufällig
**as against** gegenüber (Vergleich)
**for this reason** aus diesem Grunde
**by the way of trial** versuchsweise
**be it ever so little** wenn es auc h noch so wenig ist
**in all likelihood (or: probability)** aller Wahrscheinlichkeit nach
**it need hardly be said** es braucht kaum gesagt werden

**on the score** in dieser Beziehung
**in a manner of speaking** sozusagen
**at worst** schlimmstenfalls
**right the other way** gerade umgekehrt
**ultimately** schliesslich
**now as before** nach wie vor
**all the same** trotzdem
**in whole or in part** ganz oder teilweise
**by the way** übrigens
**in turn (or: by turns)** abwechselnd
**on the contrary** im Gegenteil
**once and for all** ein für allemal
**on the sly** insgeheim
**nothing at all** überhaupt nichts
**at a stretch** in einem Zuge
**so much the better (or: all the better)** um so besser
**at best** bestenfalls
**in round terms** rundweg
**in some ways** in mancher Hinsicht
**at length** ausführlich
**in detail** ausführlich
**by all means** ganz gewiss
**at bottom** im Grunde genommen
**three days running** drei Tage hintereinander
**to talk at random** ins Blaue hineinreden
**on no account** keinesfalls
**circumstance permitting** unter Umständen
**the last but one** der Vorletzte
**primarily** in erster Linie
**at most** höchstens
**in a sense** in gewissem Sinne
**as much again** noch einmal soviel
**in short** kurz gesagt
**to some extent** einigermassen
**in three ways** in dreifacher Hinsicht
**on the one hand** auf der einen Seite
**on the other hand** auf der anderen Seite
**one too many** einer zu viel
**one too few** einer zu wenig
**one short** einer fehlt, einer zu wenig

**on the whole** im grossen (und) ganzen
**willing or unwilling** wohl oder übel
**in all** insgesamt
**salesman and engineers alike** Verkäufer und Ingenieure
gleichermassen

## Use of preposition 'at'

**at the corner** an der Ecke
**at school** in der Schule
**at work** bei der Arbeit
**at a meeting** in (oder: bei) einer Besprechung
**good at English** gut in Englisch
**at my expense** auf meine Kosten
**he is still at it** er ist immer noch dabei
**to enter at the west gate** durch das Westtor eintreten
**at your discretion** nach Ihrem Belieben
**alarmed at** beunruhigt über
**charged at** berechnet mit
**to estimate at 50** auf 50 schätzen

## Use of preposition 'on'

**on being asked** als ich (danach) gefragt wurde
**on entering** beim Eintritt
**a lecture on** ein Vortrag über
**this is on me** das geht auf meine Rechnung
**on business** geschäftlich
**on duty** im Dienst
**to levy a duty on copper** auf Kupfer einen Zoll erheben
**to operate on batteries** von Batterien speisen
**on the radio** im Radio
**on the wall** an der Wand
**on second thought(s)** bei näherer Betrachtung
**on an average** im Durchschnitt
**on the telephone** am Telefon
**on sales (or: for sales)** zum Verkauf
**on the contrary** im Gegenteil
**on principle** aus Prinzip
**on the whole** im ganzen
**on account of** wegen

# Plans and diagrams

**arrangement plan** Anordnungsplan
**building plan (or: architect's plan)** Bauplan
**centre line** Mittellinie
**cross-section** Querschnitt
**diagram** Bild, Schaubild, Diagramm
**drilled hole** Bohrloch, Bohrung
**erection schedule (or: assembly schedule)** Montageplan
**front view** Frontansicht
**intersection line (or: cutting line)** Schnittlinie
**key plan** Hauptplan, Schlüsselplan
**list of parts (or: piece list)** Stückliste
**one-line diagram (or: single-line diagram)** Übersichts(schalt)plan
**penetrations** Durchbrüche, Durchführungen
**perspective plan (or: outline plan, master plan)** Perspektivplan
**piping plan** Rohrplan
**progress schedule** Fabrikationsplan
**rear view** Rückansicht
**rectangular hole** rechteckiger Ausschnitt
**round hole** Rundloch, rundes Loch
**schedule** Liste, Plan, Programm, Zeitplan
**schematic circuit diagram** Prinzipschaltbild
**schematic diagram** Schemazeichnung
**scheme** Schema, Plan, Anordnung; sonst auch: Anlage
**sectional drawing** Schnittzeichnung
**square hole** Quadratloch, quadratisches Loch
**top view** Draufsicht
**wiring diagram** Verdrahtungsplan

# Letter openings

**Thank you very much for your immediate reply to our questions**
Wir danken Ihnen sehr für die sofortige Beantworung unserer Fragen

**We are very pleased to note from your letter of 2nd May…**
Wir freuen uns, aus Ihrem Brief vom 2. Mai zu erfahren…

**Thank you very much for your letter of 20th May regarding…**
Vielen Dank für Ihren Brief vom 20. Mai betreffend…

**In reply to your letter of 3rd April, we have pleasure in offering you…**
Wir freuen uns, Ihnen gemäss Brief vom 3. April… anbieten zu können

**We regret to learn from your report…**
Wir beauern, Ihrem Bericht entnehmen zu müssen…

**In response to your request of 19th May, I would advise you…**
Entsprechend Ihrer Anfrage (Ihres Gesuches) möchte ich Ihnen mitteilen (auch: empfehlen, raten)…

**I am delighted to tell you…**
Es freut mich, Ihnen mitteilen zu können…

**This is to let you know…**
Hiermit möchten wir Sie informieren…

**I am sorry to have to tell you…**
Es tut mir leid, Ihnen mitteilen zu müssen…

**I must write you a few words of thanks for…**
Ich muss Ihnen ein paar Worte des Dankes wegen (für)… schreiben

**Thanking you for your above mentioned enquiry, we take pleasure in quoting you for…**
Wir danken Ihnen für die oben erwähnte Anfrage und möchten Ihnen folgendes Angebot unterbreiten:…

**In accordance with your request, we submit the following quotation:…**
Gemäss Ihrer Anfrage übersenden wir folgende Offerte:…

**We have much pleasure in attaching…**
Es freut uns, Ihnen als Beilage… übersenden zu können

**Please submit additional data providing the background information and parameters which went into this calculation**
Übersenden Sie bitte weitere grundsätzliche Angaben und Parameter, die dieser Berechnung zugrunde gelegt worden sind

**This letter confirms our verbal acceptance given to you on the 5th May at your offices**
Mit diesem Brief bestätigen wir unsere mündliche Zusage vom 5. Mai in Ihrem Hause

**Following a discussion with your (esteemed) Mr Jones on ... at ...**
Gemäss einer Diskussion mit Ihrem Herrn Jones am ... in ...

**It is understood that you have been favoured with an order for (the delivery of) two 1200 MW generators**
Uns wurde bekannt, dass Sie den Auftrag für zwei 1200-MW-Generatoren erhalten haben

# Letter endings

**We trust that we have been of some assistance (or: have been able to help you) in this matter**
Wir hoffen, das wir Ihnen in dieser Angelegenheit etwas Unterstützung geben konnten (oder; behilflich sein konnten)

**If you require any further information, please do not hesitate to contact us (or: should you require any further information on this matter, please let us know)**
Falls Sie weitere Angaben (oder: Unterlagen) benötigen, so setzen Sie sich bitte mit uns (unverzüglich) in Verbindung

**We shall be glad to supply any further information which may be required**
Gern übersenden wir Ihnen weitere Angaben (oder: Unterlagen), falls erforderlich

**Thanking you in anticipation for ...**
Vielen Dank im voraus für ...

**We should be glad if you would accept our proposals and hope to hear from you soon**
Wir würden uns freuen, wenn Sie unsere Vorschläge annehmen würden und hoffen bald wieder von Ihnen (etwas) zu hören

**We should be very obliged if you could furnish us the...**
Wir wären Ihnen sehr dankbar, wenn Sie uns die... übersenden könnten

**We trust that our explanation will cinvince you...**
Wir hoffen, dass unsere Erklärung Sie (davon) überzeugt...

**We shall let you know the result of this affair**
Wir werden Sie über das Ergebnis (dieser Angelegenheit) unterrichten

**We look forward to hearing from you again**
Wir hören gerne wieder von Ihnen

**Now we have solved our fly-ash disposal problem.**
**(Power, New York)**

# *Role readings*

## Part 1

### The failure of a circuit-breaker requires investigations by a switchgear expert

C =   Canadian customer
S =   Swiss contractor

C1:   **Is this the breaker which switches off every now and then without any apparent reason?**
Ist dies der Schalter, der sich manchmal ohne offensichtlichen Grund abschaltet?

S1:   **Yes, this is the one. It's the 3.5 kV circuit-breaker for the No. 1 forced draught fan.**
Ja, dies ist er. Es ist der 3,5-kV-Leistungsschalter für das Kesselgebläse Nr. 1.

C2:   **What do you think could be the cause of the failure?**
Was meinen Sie, was könnte die Ursache für den Ausfall sein?

S2:   **Well, I suppose it's the insufficient setting range.**
Hm, ich nehme an, der Einstellbereich reicht nicht aus (oder: ist nicht gross genug).

S3:   **But we should call in an expert.**
Wir sollten (aber) einen Fachmann heranziehen (oder herbeiholen).

C3:   **Could you arrange it to be checked by Friday?**
Könnten Sie veranlassen, dass die Überprüfung bis Freitag vorgenommen worden (oder: fertig) ist?

S4:   **Of course, Mr. Randall. Our erecting engineer will have settled this problem by Friday at the latest.**
Selbstverständlich, Mr. Randall. Unser Montage-Ingenieur (oder: Monteur) wird das Problem bis Freitag gelöst (oder: in Ordnung gebracht) haben.

**C4:**   **Could your erecting engineer check the other 3.5 kV breakers as well?**
Könnte Ihr Montage-Ingenieur auch gleich die anderen 3,5-kV-Schalter überprüfen?

**S4:**   **Most certainly. If I know Mr. Zumstein, our erecting engineer, he'll check the other breakers of his own accord. But I'll have a word with him to make sure.**
Ganz bestimmt. Aber wie ich Herrn Zumstein kenne, das ist unser Montage-Ingenieur, wird er die anderen Schalter ohnehin prüfen. Um sicher zu gehen, werde ich aber noch mit ihm sprechen.

**C5:**   **Have you ever had failures on this tpye of breaker before?**
Haben Sie (überhaupt) schon mal Ausfälle bei dieser Art von Schaltern gehabt?

**S5:**   **Not on this kind.**
Bei dieser Art noch nicht.

**Alternatively used expressions**

C2:   What do you consider...
Any idea what could...
S2:   I suspect...
I could imagine...
C3:   Could you have it checked by Friday?
S4:   Yes surely.

# Part 2

## The consulting engineer wants to visit a certain department of a factory

C =   Canadian customer
S =   Swiss contractor

S1:   **Good morning, Mr. White. I'm (Mr.) Fischer. We were expecting you earlier; was your plane delayed?**
Guten Morgen, Mr. White. Ich bin (Herr) Fischer. Wir haben Sie (schon) früher erwartet; hatte Ihr Flugzeug etwas Verspätung?

C1:   **Yes, indeed, it was half an hour late.**
Das stimmt, es hatte eine halbe Stunde Verspätung.

S2:   **You've been here before, I assume?**
Sie waren schon (mal) hier, nehme ich an?

C2:   **Yes, twice. Last time I was here for two weeks.**
Ja, (schon) zweimal. Das letzte Mal war ich zwei Wochen hier.

S3:   **Would you like a cup of coffee or something to eat?**
Möchten Sie eine Tasse Kaffee oder etwas zu essen?

C3:   **I could do with a cup of coffee (please).**
Eine Tasse Kaffee würde mir guttun.

C4:   **To start with, I would like to visit the motor factory.**
Zuerst würde ich gern die Motorenfabrik besuchen (oder: besichtigen).

S4:   **OK, if you agree we can go immediately.**
In Ordnung, wenn es Ihnen recht ist, können wir gleich gehen.

C5:   **How has production progressed since my last visit?**
Wie sind Sie mit der Fertigung seit meinem letzten Besuch vorangekommen?

S5:   **All machining has been completed.**
Die Bearbeitung aller Maschinen(teile) ist abgeschlossen.

S6:   **When will the units be assembled?**
Wann werden die Maschinen (oder: Einheiten) zusammengebaut?

**C6:** **The units are being assembled now.**
Die Maschinen werden zurzeit (oder: jetzt) zusammengebaut (oder: montiert).

## Alternatively used expressions

**S1:** Was your plane a little late?

**S3:** I could certainly use a cup of coffee, please.
(Note: This is colloquial but often used.)

**S4:** No trouble, Mr. White. Let's go straight away and discuss the details later.

**S5:** Is production on stream? (or: Is production on schedule?)

**C6:** All machining operations are finished.

**C5:** When do you expect assembly to be finished?

**S6:** Assembly is already well advanced.

**Work in the MinProg Laboratory**

**Artist's impression of work in the laboratory of the Ministry of Progress, Pulham Down.**

# Part 3

## An inspector comments on drawings submitted for approval

The Canadian inspector voices his comments to the Swiss contractor. There are several items which are not quite in accordance with the customer's specification.

C = Canadian inspector    S = Swiss contractor

C1:  **I'm sorry, but I'm not authorized to approve the use of switches and fuses in lieu of moulded-case circuit-breakers**
Es tut mir leid, aber ich kann der Verwendung von (Trenn-) Schaltern und Sicherungen, an Stelle von gekapselten Leistungsschaltern, nicht zustimmen (oder: nicht genehmigen).

S1:  **This would be the first time that this arrangement did not meet our client's full approval.**
Dies wäre das erste Mal, dass dieses System (oder: diese Anordnung) nicht die volle Zustimmung unseres Kunden findet.

C2:  **Our specification definitely calls for moulded-case circuit-breakers. Please have a look at paragraph 14.**
Unsere Bauvorschrift (oder: Spezifikation) verlangt ganz eindeutig gekapselte Leistungsschalter. Sehen Sie bitte hier, Abschnitt 14.

S2:  **Do you think that, if we offered a price reduction, we would be allowed to employ our standard switch/fuse installation?**
Was meinen Sie, wenn wir einen Minderpreis anbieten würden, könnten wir dann unsere Standard-Schalter/Sicherung-Ausführung verwenden?

C3:  **It's not a matter of costs, we simply believe that our proposed system is better in every respect.**
Das ist keine Kostenfrage, wir glauben einfach, dass unser geplantes System in jeder Hinsicht besser ist.

S3:  **What are, in your opinion, the advantages of the moulded-case circuit-breaker concept?**
Welches sind, nach Ihrer Meinung, die Vorteile des Konzepts mit den gekapselten Leistungsschaltern?

**C4:**  **There are quite a few advantages: Single encapsulated unit, all protective devices included, spare part storage simplified, only four basic types of breaker.**
Da sind einige Vorteile: Alle Teile in einem gekapselten Gerät, alle Schutzelemente sind eingeschlossen, vereinfachte Lagerhaltung, nur vier Schalter-Grundtypen.

## Alternatively used sentences

**C1:**  I'm sorry, but I cannot accept these switches and fuses instead of the specified moulded-case circuit-breakers without permission from our chief inspector (or: client).

**S1:**  This equipment has always met with our clients' full approval.

**S2:**  Yes, it's true. But our system has proved to be very reliable and is a technically sound solution.

**C3:**  That may well be, but it does represent an exception to the specification and all exceptions have to be submitted to our head office with a request for approval.

## Part 4

## Discussing results of a load analysis

An American customer and a Swiss contractor are discussing their
different calculation results concerning the installation of a 900 kW
diesel generator set.

A = American customer
S = Swiss contractor

A1: **Mr. Steinmann, do you really think we only need a 900 kW
diesel generator set?**
Herr Steinmann, glauben Sie wirklich, dass wir nur einen 900-
kW-Diesel-Generator-Satz benötigen?

S1: **Well, Mr. Whitney, our load analysis clearly shows this.**
Naja, Mr. Whitney, unsere Energiebilanz zeigt das (nunmal)
eindeutig.

A2: **You've used another approach to ours; I can now see where the
difference in results comes from. There are other load factors
involved. Look here.**
Sie haben eine andere Methode gewählt als wir; jetzt kann ich
auch sehen, wo die verschiedenen Ergebnisse herrühren. Es
sind andere Belastungsfaktoren eingesetzt worden. Sehen Sie
hier.

S2: **And some smaller simultaneity factors besides.**
Und ausserdem noch einige kleinere Gleichzeitigkeitsfaktoren.

A3: **It's now quite obvious that, according to our calculation
method, a 1000 kW set would be required.**
Es ist somit ganz offensichtlich; gemäss unserer Berechnung
würden wir ein 1000 kW-Aggregat benötigen.

S3: **We have experience with a similar emergency unit for a
chemical plant.**
Wir haben Erfahrung mit einem ähnlichen Notaggregat für eine
Chemiefabrik.

A4: **I must discuss both analyses with our chief engineer.**
Ich muss beide Analysen mit unserem Chefingenieur
besprechen.

S4:   **I would like to emphasize that a simultaneity factor of 0.4 would be sufficient for the five ‹Y4› acid pumps. We had a similar problem two years ago.**
Ich möchte betonen, dass ein Gleichzeitigkeitsfaktor von 0.4 für die fünf «Y4»-Säurepumpen ausreichen sollte. Ein ähnliches Problem hatten wir vor zwei Jahren.

**Alternatively used sentences**

A1:   Mr. Steinmann, are you sure that a diesel-generator set of 900 kW would be large enough?

S1:   Well, Mr. Whitney, the results of our load analysis show without any doubt that 900 kW will be sufficient.

A2:   Of course, you've used a different method of calculation to ours;...

A3:   I now understand why the results of our calculation indicate that a 1000 kW set would be required.

S3:   We have an almost identical set running in a chemical plant.

S4:   I would like to stress that...
Please do not forget that...

'How is the room temperature now, Miss Forsyte?'

# Part 5

## Attending works tests

**A consulting engineer is met by the chief engineer to witness the final tests on a number of vacuum circuit-breakers**

C = Consulting engineer
T = Test engineer


C1: **Can we begin with the vacuum breaker tests?**
Können wir mit den Vakuumschalter-Prüfungen beginnen?

T1: **Sorry, not just yet, we'll have to wait until the oscilloscope is ready for operation.**
Leider sind wir noch nicht ganz so weit; wir müssen warten bis das Oszilloskop betriebsbereit ist.

C2: **How long do you think it'll take you to prepare the oscilloscope?**
Was meinen Sie, wie lange wird es dauern bis das Oszilloskop aufgebaut ist?

T2: **Just ten minutes.**
Nur zehn Minuten.

C3: **There are only these vacuum breakers for testing, aren't there?**
Es sind nur diese Vakuumschalter zu prüfen, nicht wahr?

T3: **That's right. Our first line of vacuum contactors is still in the development stage.**
Ganz richtig. Unsere erste Vakuumschützen-Reihe befindet sich noch im Entwicklungsstadium.

C4: **These are the first vacuum breakers to be tested by an external inspecting authority, aren't they?**
Dieses sind die ersten Vakuumschalter, die von einer externen Prüfstelle getestet werden, nicht wahr?

T4: **Yes, they are. – Mr. White. We're ready now.**
Das stimmt. – Mr. White. Wir sind jetzt fertig (oder: bereit).

C5: **Alright. You may start the switching test.**
In Ordnung. Sie können mit den Schalterprüfungen beginnen.

T5:    **Ready! Switch on! – Now continue with the motors on full
       load.**
       Fertig! Einschalten! – Jetzt weitermachen mit Vollast.

**Alternatively used sentences**

C1:    Are you ready to begin testing?
T1:    Not quite, we're just preparing the oscilloscope.
C2:    How long will that take?
C3:    There are no contactors, are there?
C5:    Good, then let's get started. Please begin.

**Brilliant people at the MinProg Secret Service Division in Pulham
Down are proud to show their latest major achievements.**

# Part 6

## Preliminary work on obtaining a major order

Two senior engineers discuss the chances of obtaining a large order worth half a million pounds. The scope of the supply would include the delivery and installation of 3000 A circuit-breakers.

H =    Head of department
S =    Section leader


**H1:**   **We can meet the Canadian specification on all points but one.**
Wir erfüllen die (Bedingungen der) kanadischen Spezifikation in allen Punkten, einer ausgenommen.

**S1:**   **Which point do you mean?**
Welchen Punkt meinen Sie?

**H2:**   **We don't manufacture 3000 A circuit-breakers.**
Wir fertigen keine 3000-A-Leistungsschalter.

**S2:**   **No problem at all. We're going to offer breakers rated at 2400 A with forced cooling.**
Überhaupt kein Problem. Wir werden 2400-A-Schalter mit Zwangskühlung (d.h. Ventilator) anbieten.

**H3:**   **Well, it's a way round the problem.**
Na ja, damit umgehen wir das (eigentliche) Problem.

**S3:**   **We had a similar case three years ago.**
Vor drei Jahren hatten wir einen ähnlichen Fall.

**H4:**   **Ah yes, I remember, it was the contract for the two rolling mills in Johannesburg.**
Ach ja, ich erinnere mich, es war der Auftrag für die beiden Walzwerke in Johannesburg.

**S4:**   **We are performing a thorough calculation; there's even a chance that 2400 A breakers without forced cooling will be sufficient.**
Wir führen gerade eine umfassende Berechnung durch; es besteht sogar die Möglichkeit, dass wir mit 2400-A-Schaltern ohne Zwangskühlung auskommen.

H5:    **Please carry on with this calculation, and let me know the
       results.**
       Machen Sie bitte mit der Berechnung weiter und teilen Sie mir
       das Ergebnis mit.

## Alternatively used expressions

H1:    We can meet all requirements made by the Canadian specifica-
       tion except for one.

S1:    Which one is that then?

H2:    The requirements for 3000  A breakers.

S2:    Yes, at first we also believed this point to be a problem, but we
       intend offering 2400 A breakers with forced cooling.

S4:    We are also preforming a detailed calculation to ensure that our
       offer is sound. In fact there's even a chance that 2400 A
       breakers will be suitable without forced cooling.

H5:    Well, I'm pleased that you noticed this point and that you are
       working on a counter-proposal.

# Part 7

## Background information required

**The customer in Toronto telephones his contractor in Switzerland**

C =  Canadian customer
S =  Swiss contractor


S1: **Good morning, Weber, Widmer Control Systems.**
Guten Morgen, Weber, Widmer-Steueranlagen.

C1: **Good morning, Mr. Weber. This is Evans, Gibbs & Cox, speaking from Toronto.**
Guten Morgen, Herr Weber. Hier spricht Evans, Gibbs & Cox, in Toronto.

C2: **Mr. Weber, would you please send me two copies of the revised one-line diagram ME 12.05 D for the Bayswater plant as soon as possible.**
Herr Weber, würden Sie mir bitte zwei Kopien des abgeänderten (oder: revidierten) Übersichtsschaltplans ME 12.05 D für die Bayswater-Anlage übersenden, so schnell wie möglich.

S2: **I have two spare copies on my desk. I'll send them off to you this afternoon.**
Ich habe (noch) zwei Reservekopien auf meinem Schreibtisch (liegen). Ich schicke sie heute abend (an Sie) ab.

C3: **Have (or: has) the data for the current transformers been added?**
Wurden die Daten für die Stromwandler (auch) eingetragen (oder: zugefügt)?

S3: **Yes, they've been (or: it has been) added.**
Ja, sie sind (auch) eingetragen worden.

C4: **Good. I'd like to ask another favour of you. Could you send me the magnetization curves for the current transformers, also in duplicate.**
Gut. Ich hätte da noch eine Bitte. Könnten Sie mir die Magnetisierungskurven mitsenden, auch zweifach?

**S4:**  **No problem. I'll have two copies collected from the test department and will enclose them with the diagrams.**
Kein Problem. Ich lasse zwei Kopien vom Versuchslokal kommen und schicke sie zusammen mit den Plänen ab.

**C5:**  **Well, that's all for the moment. Thank you very much for your
cooperation. Goodbye, Mr. Weber.**
Das wär's im Moment. Vielen Dank für Ihre Unterstützung.
Auf Wiedersehen, Herr Weber.

**S5:**  **Goodbye, Mr. Evans.**
Auf Wiedersehen, Mr. Evans.

## Alternatively used expressions

**S2:**  I'm working on the diagram just now.
We mailed (or: posted) copies to you this morning.

**C3:**  Have you had the data for the current transformers added?

**C4:**  Send them along (or: together) with the diagrams.

**C5:**  I appreciate your immediate action. Thanks for your quick
service.

**S5:**  You're welcome. That's quite all right. It's a pleasure.

**Just order your own copy of
'Der Elektroniker'**

# Part 8

## A non-standard design

**A contractor asks his subcontractor to furnish him with a relay of non-standard setting ranges**

C =   Contractor
S =   Subcontractor


C1:   **I've learned from your Mr. Paslee that you do not have the GX-relays with the extended setting ranges on stock.**
Ich habe von Ihrem Mr. Paslee erfahren, dass Sie die GX-Relais mit den erweiterten Einstellbereichen nicht auf Lager haben.

S1:   **That's right, we do not stock this type.**
Das stimmt, wir halten diesen Typ nicht auf Lager.

S2:   **How long would it take to receive 80 of these relays, from the moment of ordering?**
Wie lange würde es dauern, bei der Bestellung von 80 Stück, bis wir diese geliefert bekommen, gerechnet vom Moment der Bestellung an.

C2:   **You could have them in eight weeks.**
Sie würden sie innerhalb von acht Wochen erhalten.

C3:   **We need them in six weeks' time at the latest.**
Wir brauchen sie in spätestens sechs Wochen.

S3:   **I'm really sorry but it's impossible to deliver you this large number in such a short time. However, you could have twenty in six weeks.**
Es tut mir wirklich leid, es ist jedoch unmöglich, diese grosse Anzahl in so kurzer Zeit zu liefern. Doch könnten wir zwanzig in sechs Wochen liefern.

C4:   **Well, that would be acceptable. It would of course mean delivery of the rest by the 20th March at the latest.**
Hm, das wäre annehmbar. Es würde dann natürlich bedeuten, dass der Rest bis zum 20. März spätestens zu liefern wäre.

S4:     **Alright, Mr. Carlisle. We'll meet these delivery dates.**
In Ordnung, Mr. Carlisle. Wir werden diese Lieferzeiten einhalten.

C5:     **You'll have our order by Monday.**
Sie erhalten unseren Auftrag bis Montag.

S5:     **Thank you, Mr. Watson.**
Ich danke Ihnen, Mr. Watson.

## Alternatively used sentences

C1:     Your Mr. Paslee advised me that…

S1:     … we dont't have that one on stock.

S2:     Eight weeks ex-works.

C4:     That would be an acceptable compromise.

C5:     Our written order will be in your hands by Monday.

# Part 9

## In favour of diesel engines?

**What nobody, a few years ago, would have believed possible has now become a reality. More and more American shipping companies are installing diesel engines in lieu of turbines in their large merchant ships.**

M =  Marketing manager
D =  Head of technical department

**D1: Did you notice that nowadays American shipping companies prefer diesel engines to turbines for ship propulsion?**
Ist Ihnen (schon) aufgefallen, dass heutzutage amerikanische Reeder Dieselmotoren den Turbinen als Schiffsantrieb den Vorzug geben?

**M1: Yes I did. But it's too early to recognize it as a definite trend.**
Ja, das ist mir aufgefallen. Es ist jedoch zu früh, hieraus einen definitiven Trend abzulesen (oder: zu erkennen).

**D2: Don't you think we should prepare a comprehensive market study?**
Sollten wir nicht eine umfangreiche Marktstudie vornehmen, was meinen Sie?

**M2: We are already working on it.**
Wir arbeiten schon daran.

**D3: Glad to hear it.**
Freut mich zu hören.

**M 3: We started a detailed survey four weeks ago.**
(Please note: Do not say 'We've started a detailed...') Vor vier Wochen haben wir mit einer eingehenden Untersuchung begonnen.

**D 4: I wonder wheter the reason for this change ist because of fuel saving measures—or because of the availability of engine diagnostic systems.**
Mich würde interessieren, ob diese Wandlung mit dem Energiesparen zu tun hat – oder weil es (jetzt) Diagnosesysteme für Dieselmotoren gibt.

**M4: Introducing the diesel diagnostic systems means saving manpower; this is indeed an important point.**
Die Einführung der Diagnosesysteme bedeutet Einsparung von Arbeitskräften; dies ist schon ein wichtiger Punkt.

**D5: Yes, the engines only have to be dismantled when the monitoring system clearly indicates the need to do so.**
Ja, (das stimmt), die Motoren sind nur dann auseinanderzunehmen, wenn das Überwachungssystem die Notwendigkeit eindeutig anzeigt.

**M5: This piston ring wear monitoring system is really a great asset for ship engine maintenance.**
Dieses Überwachungssystem für den Verschleiss von Kolbenringen ist wirklich ein grosser Gewinn für die Wartung von Dieselmotoren.

**D6: We shall look forward to hearing the outcome of your survey with much interest.**
Wir sehen mit grossem Interesse dem Ergebnis Ihrer Untersuchung entgegen (oder: Wir sind sehr gespannt auf...).

**Alternatively used expressions**

D1: Did you notice the recent swing from turbines to diesel engines the American shipping companies are showing for ship propulsion.

M1: Yes, I noticed it some time ago. It's difficult to judge whether it will hold or not.

M2: I'm pleased you see the necessity, we are already hard at it.

**How to tackle the big problems—that's what you'll learn at the Ministry of Progress, Pulham Down**

# British and American English—
# two languages?

Sometimes the question arises 'Are there really so many differences between British and American English?' Somme years ago, I tried to find the answer to this question by carrying out an experiment involving British and American engineers.

I handed over an American text of about 20 lines to five British engineers independent of each other. Asking them, if it was a British text, they all told me, fully convinced, that it was a genuine British English passage. Then I gave a British text to each of five American engineers, and they all told me it was definitely a text in American English. However, I must admit, I had taken care that these texts did not include obvious British and American spellings. This restriction was necessary to make the experiment possible.

The two varieties of English have never been so different as people have imagined, and the dominant tendency, for several decades now, has clearly been that of convergence and even greater similarity.

*Georg Moellerke*
El.-Ing. und staatl. gepr. Übersetzer

# *Translation exercises*

**A  Singular or plural**

a) Hier sind die Daten (oder: Unterlagen), wonach du mich gestern gefragt hast.
b) Diese Geräte sind wirklich nicht teuer.
c) Diese Informationen sind der «Elektrotechnik» entnommen.
d) Unsere Kenntnisse über das Plasmaverhalten sind immer noch begrenzt.
e) Die Fortschritte, die wir im Schaltungsaufbau gemacht haben, sind wirklich bemerkenswert.
f) Die Arbeiten am Kernkraftwerk Leibstadt gehen immer noch weiter.
g) Man hat uns mitgeteilt, dass der Inhalt von einigen Kisten beschädigt wurde.
h) Deine Schutzbrille ist auf der Werkbank.
i) Die Ware ist bereits verkauft worden.
j) Die zum Labor führende Treppe ist sehr steil.
k) Diese Nachrichten sind zu gut, um wahr zu sein.
l) Die Vereinigten Staaten sind auf dem Gebiet der Elektronik führend.
m) Die Ingenieure haben ihr Bestes getan.
n) 500 Dollar sind ein guter Preis für dieses gebrauchte Oszilloskop.

**B  With or without article**

a) So ist nun mal das Leben eines Elektronikers; er muss sich laufend für neue Entwicklungen interessieren.
b) Die Stelle des Gruppenleiters in der Vertriebsabteilung ist frei.
c) Die Sitzung ist gerade zu Ende gegangen.
d) Die Zeit ist auf unserer Seite.
e) Wir arbeiten seit zehn Stunden ohne Unterbrechung.
f) In der Regel sollte man alle Verbindungen überprüfen, bevor man mit den Messungen beginnt.
g) Als Maschinenbau-Ingenieur sollten Sie mir einen guten Rat geben können.
h) Was für eine Farbe hat das neue Steuerpult?

## C  The adverb

a) Er hat die Messungen schnell und gründlich erledigt (oder: durchgeführt).

b) John ist kein schneller Konstrukteur, aber er arbeitet sorgfältig.

c) Diese Steuerungsanlage ist hydraulisch und arbeitet mittels Mikroprozessor völlig automatisch.

d) Der Metronik-Mikroprozessor ist überraschend leistungsfähig.

e) Offensichtlich war die Batterie völlig leer (oder: entladen).

f) Der Ingenieur (dort) spricht wahrscheinlich auch Englisch.

g) Wenn Sie ihm keine Gehaltserhöhung geben, wird er schnell das Interesse verlieren.

h) Ich kenne den Prüffeldleiter nicht persönlich.

i) Wir erhalten häufig Aufträge aus dem Ausland.

## D  Auxiliary verbs

a) Ich kann die Analyse selbst vornehmen (oder: durchführen).

b) Können Sie mir Ihr Oszilloskop leihen?

c) Kann ich die Anlage jetzt in Betrieb setzen?

d) Die Sicherung kann jeden Moment durchbrennen.

e) Der Chefkonstrukteur kann diesmal recht haben.

f) Unser Mikroprozessor konnte nicht (oder: war nicht imstande), auf diesen Spannungsabfall zu reagieren.

g) Konnten Sie mit dem Abteilungsleiter sprechen?

h) Soll ich das Steuergerät jetzt nachstellen?

i) Warum will denn der Schichtleiter nicht kommen?

j) Ich hätte den Rat meines Schaltungsfachmanns befolgen sollen.

k) Sie brauchen das Instrument nicht zu nehmen, wenn Sie nicht wollen.

l) John hätte dieses Problem lösen können.

## E  The passive

a) Diese Schalter werden überall verkauft.

b) Er wurde bei einem Unfall im Versuchslokal verletzt.

c) Ist der Schaden der Versicherung gemeldet worden?

d) Diese Maschine kann von Hand oder automatisch gesteuert werden.

e) Die neuesten Zahlen werden nächste Woche veröffentlicht.

f) Das Angebot sollte doch gestern abgeschickt werden, nicht wahr?

g) Das Gerät wird im Moment (oder: gerade) untersucht.
h) Von wem wurde die Bauvorschrift übersetzt?
i) Von wem wird die Rechnung bezahlt werden?
j) Wodurch wurde der Schaden verursacht?
k) An wen wurde der Bericht geschickt?
l) Wann wird man uns die Zeichnungen schicken?
m) Man wird die Presse informieren müssen.

### F  Conditional sentences

a) Es wäre nützlich, wenn der Schaltungsfachmann auch hier wäre.
b) Wenn ich Sie wäre, würde ich mich beim Pressereferenten beschweren.
c) Wenn es nicht einen Kurzschluss gegeben hätte, wäre das alles nicht passiert.
d) Wenn das Flugzeug pünktlich gewesen wäre, hätte ich den Bus nicht verpasst.
e) Wenn der Elektromonteur die Abänderung in der Schaltung gekannt hätte, wäre er nicht so unbesorgt gewesen.
f) Wenn die Pressevorführung am Mittwoch gewesen wäre, wären mehr Journalisten gekommen.
g) Wenn ich mehr Geld hätte, würde ich auch ein grösseres Laboratorium bauen.

### G  Present continuous and simple present

a) Wir machten gerade eine Pause, als das Telex eintraf.
b) Ich nehme meistens den Pendelwagen der Firma.
c) Ich lese im Moment die «Elektrotechnische Rundschau».
d) Die Möglichkeit, das Gerät auszuwechseln, wird immer noch geprüft.
e) Dieses Experiment beweist viele Dinge.
f) Was versuchen Sie zu beweisen?
g) Jetzt verstehe ich, was Sie damit meinen.
h) Wie denkst du über den Vorschlag, die Schaltung abzuändern?
i) Wann treffen Sie den Abteilungsleiter?
j) Wie lange bleiben die Maschinenschlosser bei uns?
k) Das Lager macht in ein paar Minuten zu.
l) Alan Jones übernimmt die Entwicklungsabteilung im Januar.

**H  Do(n't)—does(n't)—did(n't)**

a) Willst du den Zeichnern nicht helfen?
b) Er brauchte das Oszilloskop nicht.
c) Sind Sie nicht Amerikaner?
d) Was bieten Sie normalerweise Ihren Kunden zum Essen an?
e) Mussten Sie die Analyse selbst vornehmen?
f) Diese Art von Firma hat kein eigenes Labor.
g) Haben Sie oft Streiks in Kanada?
h) Sei doch nicht so ungeduldig, die Aufstellung (oder: Liste) ist gleich fertig.
i) Wer sah ihn auf dem Prüffeld?
j) Wen sah er im Labor?
k) Wem gab er das Werkzeug?
l) Was tat er, um den Elektromonteuren zu helfen?
m) Was machte den Prüffeldleiter misstrauisch?
n) Wer will die Vorführung nicht sehen?

**I  Perfect or present tense?**

a) Sind Sie schon mal im amerikanischen Stammhaus gewesen?
b) Ich bin gerade aus Washington zurückgekommen.
c) Ich bin noch nie im New Yorker Zweigbüro gewesen.
d) Wie hat Ihnen San Francisco gefallen?
e) Wie lange kennen Sie Mr. Watson schon?
f) Oh, ich kenne ihn seit mindestens zehn Jahren.
g) Sagen Sie, wie lange sitzen wir hier (eigentlich) schon?
h) Ich glaube, wir sind gegen halb vier gekommen.
i) Dann warten wir jetzt seit zwanzig Minuten auf unseren Bericht.

# Key to translations

**A**

a) Here are the data you asked me for yesterday.
b) This equipment is (or: these devices are) really not expensive.
c) This information has been taken from ‹Electrical Engineering›.
d) Our knowledge of plasma behaviour is still limited.
e) The progress we have made in circuit design is really (or: quite) remarkable.
f) The work on Leibstadt nuclear power station ist still going on.
g) We have been informed that the contents of some cases were damaged.
h) Your safety glasses are on the workbench.
i) The goods have (or: the merchandise has) already been sold.
j) The stairs leading to the laboratory are rather steep.
k) This news is too good to be true.
l) The United States are leading in the electronics field.
m) The engineers have done their best.
n) 500 Dollars is a good price for this second-hand (or: used) oscilloscope.

**B**

a) Such is life for an electronics engineer; he has to keep up continuously with the latest designs (or: developments).
b) The position of section leader in the sales departement is vacant.
c) The meeting has just come to an end (or: has just ended).
d) Time is on our side.
e) We have been working for ten hours without a break.
f) As a rule, one should check all connections before starting the measurements (or: before starting to take readings).
g) As a mechanical engineer you should be able to give me some good advice.
h) What colour is the new control console (or: control desk)?

**C**

a) He has done (or: he made) the measurements quickly and thoroughly (or: he took the readings…).
b) John is not a quick working draughtsman, but be works carefully.
c) This control equipment is hydraulic and works fully automatically by means of a microprozessor.

d) The Metronik microprocessor is surprisingly efficient.
e) Obviously the battery was completely discharged.
f) That engineer probably speaks English, too.
g) If you don't give him a pay rise (or: rise in salary), he will quickly loose interest.
h) I don't know the chief test engineer personally.
i) We frequently receive orders from abroad.

**D**
a) I can carry out (or: do) the analysis myself.
b) Can you lend me your oscilloscope?
c) Can I (or: may I) put the system into operation now?
d) The fuse can blow (at) any moment.
e) The chief designer may be right this time.
f) Our microprocessor could not respond to this voltage drop.
g) Were you able to speak to the departmental head (or: head of department)?
h) Shall I reset (or: readjust) the controller now?
i) Why doesn't the shift engineer want to come?
j) I should have taken the advice of my circuit designer.
k) You needn't take that instrument if you don't want to.
l) John could have solved this problem.

**E**
a) These switches are sold everywhere.
b) He was injured in an accident in the test bay.
c) Has the damage been reported to the insurance company?
d) This machine can be operated by hand (or: manually) or automatically.
e) The latest figures will be published next week.
f) The offer was to be sent off yesterday, wasn't it? (or: was to have been sent off)
g) The device ist just being examined.
h) Who was this specification translated by?
i) Who will the bill be paid by?
j) What was the damage caused by?
k) Who was the report sent to?
l) When will we be sent drawings?
m) The press will have to be informed.

**F**  It would be useful if the circuit designer was here too (or: were here
a)  as well).


b)  If I were you, I would complain to the public relations officer.
c)  If there hadn't been a short-circuit, all this would not have
    happened.
d)  If the plane had been on time, I wouldn't have missed the bus.
e)  If the electrical erector had known the circuit modification he
    wouldn't have been so unconcerned.
f)  If the press performance had been on Wednesday, more journalists
    would have come.
g)  If I had more money, I too would build a larger (or: bigger)
    laboratory.


**G**
a)  We were having a break when the telex arrived.
b)  I usually take the company's shuttle car (or: van).
c)  I am reading the 'Electrical Review' at the moment.
d)  The possibility of exchanging (or: replacing) the device is still being
    investigated.
e)  This experiment proves (or: verifies) many things.
f)  What are you trying to prove?
g)  Now I understand what you mean by that.
h)  What do you think of the proposal to alter the circuitry?
i)  When are you meeting the head of department (or: departmental
    head)?
j)  How long are the (engine) fitters staying with us?
k)  The stores are closing in a few minutes.
l)  Alan Jones is taking over the development department in January.


**H**
a)  Don't you want to help the draughtsmen?
b)  He didn't need the oscilloscope.
c)  Aren't you American?
d   What do you normally offer your customers to eat?
e)  Did you have to carry out (or: perform) the analysis yourself?
f)  This kind of company doesn't have a laboratory of its own.
g)  Do you often have strikes in Canada?
h)  Don't be so impatient, the schedule will soon be ready.
i)  Who saw him at (or: in) the test bay (or: test lab)?

j)  Who did he see in the laboratory?
k)  Who did he give the tool to?
l)   What did he do to help the electrical erectors?
m) What made the chief test engineer suspicious?
n)  Who doesn't want to see the demonstration?


**I**
a)  Have you ever been to the American head office?
b)  I have just come back from Washington.
c)  I have never been to the New York subsidiary.
d)  How did you like San Francisco?
e)  How long have you known Mr. Watson?
f)  Oh, I've known him for at least ten years.
g)  Tell me, how long have we been sitting here already.
h)  I think we came at about half past three.
i)   Then we've been waiting for our report for twenty minutes.

**Novel food tests are being performed at the Ministry of Progress, Pulham Down.**

# DER TECHNIKER AUF AUSLANDSREISE, IM GESPRÄCH MIT KUNDEN UND EXPERTEN

Der ins Ausland reisende Techniker kennt meist „seine" englischen Spezialausdrücke, aber gewisse Wendungen der fachlichen Umgangssprache fallen ihm einfach nicht ein. Da wird er froh sein, wenn er sich im Flugzeug noch rasch ein wenig in dem Taschenbuch umsehen und auf die bevorstehenden Gesprächssituationen vorbereiten kann. Kernstück dieses „technischen Sprachführers" sind neun Episoden „The experiences of Bob Keller, an engineer". Auf kurzweilige Art werden hier die Erlebnisse eines Ingenieurs auf seiner Kanada-Reise geschildert.

## ZUM AUTOR

Ing. Georg Möllerke, Jahrgang 1930, war nach der Ausbildung zum Elektroingenieur elf Jahre als Konstrukteur und Projektingenieur in der Schiffselektrotechnik und anschließend im Geschäftsbereich Hochspannungs-Schaltanlagen eines Schweizer Großunternehmens tätig. Seit 1972 Redakteur, ist er heute Herausgeber und Verleger der von ihm gegründeten Zeitschrift „Engineering Report".

ISBN-13: 978-3-540-62357-1